Saving and Economic Growth:
Is the United States Really Falling Behind?

by
Robert E. Lipsey
Irving B. Kravis

*Jointly Sponsored by the
American Council of Life Insurance
and
The Conference Board*

Contents

WHY THIS REPORT	vii
INTRODUCTION	ix
1. INCOME LEVELS AND GROWTH AT HOME AND ABROAD	**1**
The Position of the United States in the World Economy	1
The Growth of Nations	5
What Is a 'Fast' Rate of Growth?	7
A Golden Age of Economic Growth	8
What Growth Rates Shall We Demand of the Economy?	10
The Relative Stability of U.S. Long-term Growth Rates	12
Within-Period Variations in Growth	14
Conclusions	19
APPENDIX 1A. Some Fine Points	20
APPENDIX 1B. Data Sources	22
Table 1-1. Per Annum Growth Rates, Population and Real GDP, Selected Countries and Groups of Countries, 1950-1984	6
Table 1-2. Historical Growth Rates for Selected Industrial Countries	9
Table 1-3. The Catch-Up of Industrial Countries Relative to U.S., Real GDP per Capita, 1950-84	14
Table 1-4. Annual Growth Rates of Total and Per Capita GDP, Selected Countries and Areas, and Selected Periods, 1950-84	16
Appendix Table 1-1. World Population and Production, Selected Countries and Groups of Countries, 1984	24
2. SAVING AND CAPITAL FORMATION IN THE UNITED STATES AND OTHER INDUSTRIAL COUNTRIES	**25**
Is the United States a Spendthrift Nation?	25
Finding a Yardstick to Measure Saving and Investment	25
Whose Saving Should be Measured?	28
What Should be Counted as Saving?	29
Consumer Durables	30
Education	31
Research and Development	32
Military Capital Formation	33

    Other Adjustments to the Investment Concept.............. 33
    Real Investment Ratios: How Much Capital Goods Does the
      Saving of Different Countries Buy?...................... 35
    Per Capita Capital Formation............................. 37
    Has Low Saving Made the United States Poor in Conventional
      Capital?.. 39
    Has the United States Misallocated Its Capital Formation?....... 41
    The Distribution of Conventional Capital Formation Among
      Types of Assets... 41
    The Distribution of Broadly-Defined Capital Formation........ 43
    The Distribution of the Stock of Capital Among Types of
      Assets.. 46
    Why Do Saving and Capital Formation Rates Differ Among
      Countries?.. 47
    Effects of Taxation...................................... 48
    Effects of Government Saving............................. 50
    Saving vs. Capital Formation............................. 52
    Summary.. 53

Table 2-1. International Comparison of Saving Rates, 1970-80
(Published Estimates without Adjustments).................... 29
Appendix Table 2-1. General Government and National Gross Saving
as Percent of GDP, Conventional Definition of Saving, OECD
Data, 14 Countries... 56

3. THE RELATION OF CAPITAL FORMATION TO ECONOMIC GROWTH................................... 57
    The Association between Capital Formation and Economic
      Growth.. 57
    From Capital Formation to Growth or From Growth to Capital
      Formation?.. 60
    Two Schools of Thought................................... 60
    Some Empirical Evidence on the Timing of Capital Formation
      and Growth... 61
    Has Slow Capital Accumulation Led to Slow U.S. Growth?....... 62
    Conclusions.. 64

Appendix Table 3-1. Coefficients for Capital Formation and Growth
of Labor Force and Labor Force Participation Rates in Equations
for Growth in GDP and GDP per Capita........................ 65
Appendix Table 3-2. Coefficients for Capital Formation Ratios in
Preceding Periods and Growth in Labor Force and Labor Force
Participation Rates in Equations for Growth in GDP and GDP
per Capita.. 66
Appendix Table 3-3. Coefficients for Capital Formation Ratios in
Following Periods and Growth of Labor Force and Labor Force

Participation Rates in Equations for Growth in GDP and GDP per Capita... 67
Appendix Table 3-4. Sources of Economic Growth in the United States and in Northwest Europe, 1950-1962..................... 68
Appendix Table 3-5. Average Annual Growth Rates of Real Product and Parts Accounted for by Growth of Real Inputs, Nine Countries, Pre-1960 and 1960-1973............................ 69

4. SUMMARY... 71
Differences in Saving and Capital Formation.................. 72
Capital Formation and Economic Growth...................... 77
Looking Ahead.. 78

LIST OF REFERENCES....................................... 83

Charts
1-1. Levels of Income, 1984.................................... 4
1-2. Catching Up to the United States.......................... 13
1-3. The Slowdown in Economic Growth, 1950-84................ 15
2-1. Capital Formation as Percent of GNP...................... 27
2-2. Capital Formation, 1980.................................. 36
2-3. Stock of Consumer Durables.............................. 39
2-4. Nonresidential Capital Stock per Person Relative to the United States... 41
2-5. Distribution of Capital Formation by Type................. 43
3-1. Relationship Between Capital Stock and Income, Seven Countries... 59

Acknowledgments

We would like to express our appreciation to the American Council of Life Insurance and The Conference Board for providing the financial support that made this study possible. We are particularly grateful for the continuing encouragement from Kenneth M. Wright of the American Council of Life Insurance and Edgar R. Fiedler of The Conference Board during the research and editorial process. Harvey H. Segal helped us with many editorial suggestions to improve the clarity of the manuscript. We also wish to thank Linda Molinari and David Robinson for statistical and programming assistance and James Hayes and Rosa Schupbach for the preparation of the manuscript.

The opinions expressed in this study are those of the authors and do not necessarily reflect the views of the American Council of Life Insurance or The Conference Board.

Irving B. Kravis
University Professor Emeritus
 of Economics
University of Pennsylvania

Robert E. Lipsey
Professor of Economics
Queens College and Graduate Center
City University of New York

Why This Report

Business leaders in the United States have long recognized the close relationship between saving and investment, on the one hand, and the pace of economic growth, on the other. In recent years, concern has been mounting that the United States is falling behind other industrial nations, where capital formation appears to exceed our own by a wide margin. The prime example is Japan, where saving and investment ratios are reported to be three times those of the United States, providing a strong base for the exceptional performance of the Japanese economy.

But do we really save and invest less of our output than other nations? Is the United States becoming capital poor? How close is the linkage between investment and economic growth? To examine these complex issues in a thorough and objective fashion, The Conference Board and the American Council of Life Insurance invited two highly regarded authorities on international economics to undertake this study. Robert E. Lipsey is Professor of Economics at Queens College and the Graduate Center of the City University of New York; Irving B. Kravis is University Professor Emeritus of Economics at the University of Pennsylvania. Both have also been long-time members of the research staff at the National Bureau of Economic Research. There and elsewhere they have collaborated on a number of authoritative studies that involve international comparisons of economic data.

We asked Professors Lipsey and Kravis to review the international data and research on these topics as a basis for reaching their own conclusions, which are interesting and provocative. The authors challenge much of the conventional wisdom on the relative performance of the U.S. economy and on the relationship between saving and growth. Their analysis will not settle these questions once and for all, but it undoubtedly makes an important contribution by clarifying the issues.

This study will, we believe, be useful to business people, economists, and government policymakers in separating fact from fiction in this area of increasing national concern. Our two organizations are grateful to the authors for shedding new light on a crucial matter.

James T. Mills, President
The Conference Board

Richard S. Schweiker, President
American Council of Life Insurance

Introduction

It has been fashionable in recent years to take a gloomy view of the international economic position of the United States. The country is depicted as a community of spendthrifts, unwilling to save for the future or to provide the tools for economic growth by building up or renewing its stock of machinery and equipment. With Japan as the model, it is believed that such investment is the key to rapid economic growth. One penalty for this supposed neglect of the future is sluggish economic growth. Closely related is the apparent catching up or approach to U.S. income levels in the late 1970s or early 1980s by other countries. There is also fear that other countries, particularly Japan, will overtake the United States in technological leadership and productivity.

Explanations abound for the seeming loss of U.S. leadership. Some argue that Americans are sharply focused on enjoying the present, in contrast to Europeans and, in particular, the Japanese, who are cautious and concerned about the future. Others contend that U.S. companies, dependent on the favor of money managers whose horizons are limited to the current quarter's profits, cannot take a long-term view. Still others point to the U.S. tax system, claiming it is biased against saving, investment, and income from capital, compared to those of other countries. Blame is also placed on Social Security or other public insurance programs for discouraging private saving.

We pose three sets of questions for those who believe that the U.S. is in economic decline:

1. Are there standards by which the U.S. growth of total output and output per capita can be judged unsatisfactory? Has the growth of output per capita in the United States slowed in comparison with past rates? Has the United States fallen behind other countries in the levels of income or output per person? What are the growth rates for output—both aggregate and per capita—that can reasonably be expected of the U.S. economy, and of other countries?

2. Is the United States improvident? Does it really save and invest far less of its output than other developed countries or is that largely an illusion, based on particular, arbitrary definitions of saving and investment? Have U.S. saving rates in the last few decades been below historical levels? Has the United States become capital poor as the result

of low saving rates? Has it been consuming rather than investing for future production? What explains the differences in saving rates among countries?

3. What is the role of saving and investment with respect to the U.S. growth rate? Does the rate of economic growth depend mainly on the proportion of a country's income devoted to building up its stock of plant and equipment? Must a high level of such investment precede a period of rapid growth, or are there other elements of equal or greater importance?

Our answers to these questions are, in the main, less pessimistic than the views often expressed. Our analysis does not raise the performance of the U.S. economy to star status, but suggests that much of the criticism of the American public for its overspending is based on misinterpretation or, at least, questionable interpretation of the data. Furthermore, much of the pessimism about economic growth rates is the result of expectations that are unrealistic when viewed in historical perspective.

In examining these issues, we usually compare the United States with other industrial countries, particularly the six that have emerged with the United States as the "summit" countries in the international arena—Japan, Canada, France, Italy, West Germany, and the United Kingdom. We include other developed countries when data are available, or reduce the number when there is a lack of comparable statistics. On some general issues, we extend our comparisons to all developed countries, or to all countries. Overall, however, the countries covered are those usually thought of as serious rivals of the United States for economic leadership or as major competitors in world markets.

Chapter 1
Income Levels and Growth at Home and Abroad

It is widely held that the position of the United States in the world economy has deteriorated over the last 30 years, and that it is no longer the unequivocal leader, even in terms of per capita income. The catching up of other countries is often noted, but what has not received much attention is the possibility that such changes stem more from the increased dynamism of partners and rivals than from a deterioration in the long-term performance of the U.S. economy.

These developments are examined in the course of this chapter, which provides the background—in terms of world production and income—for the ensuing discussion of saving and capital formation. We compare per capita income levels, as well as current and historical growth rates, for the United States and other countries, and then consider the range of growth rates that can be expected for the U.S. economy. In estimating that range, we must choose either to rely on the historical record of the United States or to adopt criteria based on the recent experience of countries with very high growth rates.

The Position of the United States in the World Economy

We begin with a brief account of the U.S. standing in the world economy at the end of 1984. More information may be found in Appendix Table 1-1.

- The United States, with five percent of the world's population, accounted for almost a fifth of world output and income.
- The 21 advanced industrial countries, including the United States, produced and absorbed nearly half the world's real output of goods and services, but contained less than one-sixth of the world's people.
- The 10-member European Community (EC) had a larger population than the United States; real output was lower, however.
- Japan had a little over half the population, but well under half the output, of the United States. Overall, Japan had a smaller population and lower real output than the EC, but ranked higher than any individual European country on both scores.
- Less developed countries (LDCs) with market economies had over half the world's population, but produced only a quarter of its real output and income.

Why Exchange Rates Produce Incorrect Conversions

The very common practice of converting values reported in different national currencies to a common currency—generally dollars—by means of exchange rates leads to seriously distorted results. It assumes that prices, in terms of dollars, are equal everywhere, or, in other words, that if two units of a currency exchange for one dollar, they will buy the same amount of goods and services in that country that a dollar will buy in the United States. Yet, as any traveler knows, some countries are expensive and others are not, and poor countries tend to be among the latter. In fact, for the basket of goods that make up GDP, the purchasing power of the currencies of very low income countries has been found to be two or three times as great as the exchange rate. That means that the use of exchange rates to convert the gross products of other countries to dollars systematically understates the real product of lower income countries. This error is avoided in Appendix Table 1-1 by using the purchasing power of currencies rather than exchange rates to convert gross products to dollars. The purchasing power estimates were derived mainly from the U.N. International Comparison Project (see Appendix).

The errors arising from conversion by exchange rates are especially troublesome in periods when rates are volatile. Obviously incompatible estimates are

- Centrally planned economies (CPEs) accounted for about 30 percent of the world's population (China alone had 22 percent) and for a lower proportion of its real output and income.

A few comments about the concepts used throughout this report:

Gross *domestic* product (GDP) rather than Gross National Product (GNP) is used here because it is the concept adopted in the United Nations standardized system of national accounts (United Nations, 1968) and is the basic measure used by most countries. The difference is that the *domestic* product includes all output within the territorial boundaries of a country whether produced by residents or non-residents, while the *national* product measures the output of residents whether they produce at home or on foreign territory. The differences are small for the United States and most other countries. In 1982, for example, the U.S. GDP was $3,025.7 billion and net factor income from abroad was $47.3 billion, adding up to GNP of $3073.0 billion, a difference of 1½ percent. Although GNP is the central concept used in the U.S. system of accounts, the United States publishes a GDP estimate and provides the United Nations with accounts based on the domestic concepts; it is those figures that are used here.

The output and income figures for gross domestic product are converted to U.S. dollars on the basis of the relative purchasing power parity (PPP) of each currency, rather than by exchange rates. The GDP of France, for example, is converted to dollars by dividing it by the number of francs required to buy a dollar's worth of GDP, rather than the number of francs required to buy a U.S.

> sometimes produced for successive periods. Unfortunately, even the widely cited *World Atlas* of the World Bank continues to rely on exchange rate conversions and thus continually falls into this trap. For example, the 1985 edition reports the following per capita GNP figures in U.S. dollars for Sweden and the U.S., which are purportedly comparable.
>
> | | 1982 | 1983 |
> |---|---|---|
> | Sweden | $13,840 | 12,400 |
> | United States | $13,160 | 14,090 |
>
> The figures imply that Sweden's per capita was 5.2 percent higher than that of the United States in 1982 and 12 percent lower than that of the United States the very next year. The improbability of such large, sudden shifts, barring a disaster in the declining country, is underscored by the fact that the national accounts data of the two countries show that per capita GDP in constant prices rose by roughly about the same percentage between 1982 and 1983 (2.3 percent in Sweden and 2.6 percent in the United States). The moral of the story is that the level and movement of exchange rates do not reflect purchasing power levels and changes. Of course, exchange rates are important parameters of the world economy, but they are not suited to the comparison of real values.

dollar on the foreign exchange market. The use of GDP rather than the more familiar GNP has no appreciable influence on the results. Conversion by PPPs, however, systematically yields higher incomes for poor countries because prices tend to be low in a poor country and the purchasing power of its currency thus tends to be greater than indicated by the exchange rate. The U.S. share of world output and income looms larger when exchange rates are used to convert the GDPs of other countries to dollars; but that result is misleading. (See box above.)

The justification for treating GDP as a measure of gross *income* and of gross *output* is that the income available to countries in real terms is the bundle of commodities and services they produce and either absorb at home or trade for foreign goods or assets. However, GDP figures should be regarded only as an approximate guide to the relative incomes of the countries, because among other limitations they include depreciation allowances—estimates of capital used up in production—that should not properly be counted as part of net income. We bring population and GDP figures together in Chart 1-1, which shows levels of per capita GDP in 1984. Real per capita GDP is the best overall indicator of economic performance, reflecting the results of whatever combination of good fortune, human wisdom, and effort produces economic success.

Although in 1984 the United States had the highest real per capita GDP of the countries charted, we cannot be sure it had the world's highest income. For countries with very similar per capita incomes, such as the United States and Canada, the margin of error could well reverse the ranking order. In 1980, for example, the data source used for the chart puts Canada ahead of the United

Chart 1-1
Levels of Income, 1984
(Real Gross Domestic Product per Capita)
U.S. = 100

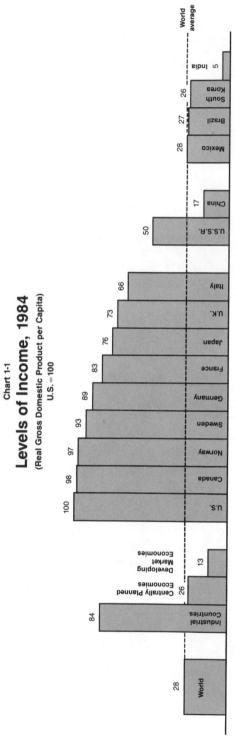

Economic Growth, 1950-1984
(Real Gross Domestic Product per Capita, Annual Percent Change)

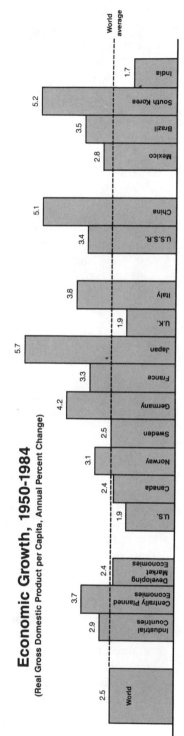

Source: Table 1-1 and Appendix Table 1-1.

States. The most we can say with confidence is that the United States had one of the highest per capita incomes—perhaps the highest—in the world, but it was no longer far ahead of the nearest runner-up. In addition to the United States and Canada, the top countries in terms of per capita income include Norway, Sweden and Switzerland. France and Germany were next, followed by Japan, the United Kingdom and Italy. Japan, for all its remarkable competitiveness in world markets, had only a little over three quarters of the real U.S. per capita. It had forged ahead of Italy and the United Kingdom, but its per capita income was less than that of Germany and France.

There is a wide variety of income levels among developing countries with market economies. Most have low incomes, and even the relatively better off countries of the group, such as Brazil, have per capita incomes of 11 to 27 percent of the U.S. level.

Although the figures on real GDP eliminate the gross distortions resulting from the use of exchange rates, they are still subject to some degree of error because of the difficulties encountered in international price comparisons and the incomparabilities that are inevitable in the underlying national GDP estimates compiled by different teams of statisticians. In our judgment, great weight should not be given to differences in per capita income of less than five percent between high income countries.

The Growth of Nations

The figures in the upper half of Chart 1-1 represent a snapshot of a given period of time, and their interpretation is affected by the rates of change used to calculate them. For example, in 1984, Japan's per capita output was 76 percent of that of the United States; for France the figure was 83 percent. The data take on different significance in the light of the very different growth rates of the three countries, however. Japan, captured in the still-picture of Chart 1-1, is like a speedboat gaining on France and the United States, between which the distance is narrowing slowly and erratically.

The growth record for aggregate GDP, population, and GDP per capita for the period from 1950 to 1984 is set out in Table 1-1 for a number of countries and country groups. We focus on this 34-year period, which begins after much of the damage and dislocation of World War II had been offset. The facts may be summarized briefly.

With respect to total real GDP:

- The United States grew slowly in relation to other countries. Its 3.25 percent yearly growth rate was lower than that of any country shown, except for the United Kingdom and Sweden; it was less than half the rate of fast-growing Japan.
- The world growth rate of 4.45 percent per annum was well above the U.S. rate. Averages for each of the three major groups of countries—industrial, developing, and centrally planned—also exceeded the U.S. rate.

Table 1-1: Per Annum Growth Rates, Population and Real GDP, Selected Countries and Groups of Countries, 1950-1984

	Per Annum Growth Rate of		
	Population	Real GDP	Real GDP per Capita
Industrial Countries (21)	.97	3.86	2.86
U.S.	1.31	3.25	1.91
Canada	1.80	4.25	2.41
European Community (10)	.59	3.75	3.14
France	.80	4.16	3.33
Germany	.60	4.81	4.19
Italy	.56	4.33	3.75
U.K.	.31	2.26	1.94
Norway	.71	3.83	3.10
Sweden	.50	3.01	2.50
Japan	1.10	6.88	5.73
Developing Market Economies (86)	2.39	4.82	2.37
Low Income Countries (29)	2.27	3.70	1.40
India	2.17	3.89	1.68
Middle Income Countries (57)	2.52	5.21	2.62
Oil Exporters (6)	2.46	5.42	2.89
Relatively Industrialized (17)	2.42	5.16	2.68
Brazil	2.71	6.33	3.53
Korea	2.05	7.32	5.17
Mexico	3.09	5.98	2.81
Other (34)	2.72	5.16	2.37
Centrally Planned Economies (8)	1.63	5.41	3.72
China	1.87	7.07	5.11
USSR	1.25	4.66	3.36
World (115)	1.87	4.45	2.53

Source: See appendix.
Per annum growth rates between terminal years.
The number of countries included in each group is shown in parentheses.

- Centrally planned economies registered the fastest growth, the developing countries were in a middle position, and the industrial countries were the slowest growing group.
- Although the developing countries are not our primary concern, it is worth noting that the high average rate of growth covers a diversity of experience, with some countries performing closer to the low U.S. rate and others approaching the high Japanese rate.

The story of per capita growth rates portrayed in the lower half of Chart 1-1 reflects the effects of differences in population growth rates. The growth of the U.S. population, at 1.31 percent per annum, was high relative to most industrial countries and low relative to developing countries. Consequently, the gap between the U.S. growth rate and that of the industrial countries was larger on a per capita basis. The U.S. real GDP per capita growth rate of 1.91 percent compares with an industrial country average of 2.86 percent. At the same time, the high rates of population growth in developing countries (2.39 percent per annum) lower their per capita growth rate to 2.37 percent, which is much closer to the U.S. rate.

The ranking of countries with respect to growth rates probably deserves more credence than differences in magnitude. Reasons to be skeptical about the comparability of the growth estimates have already been given and are elaborated in Appendix 1A (see pages 20-21). We hazard the opinion that differences in growth rates smaller than 0.5 percentage points between two developed countries should not be given much weight. That may also be true of differences of less than one percent per year between a developed country and a less developed one. By this standard, one can delineate three groups of industrial countries in terms of growth of real per capita GDP based on the data in Table 1-1:

Low	United States, United Kingdom, Canada, Sweden
Middle	France, Germany, Italy, Norway
Top	Japan

How shall we assess the record summarized in Table 1-1 for the world economy? For the United States? It demonstrates world GDP expanding at 4.5 percent per annum in the aggregate and 2.5 percent on a per capita basis. Although there is considerable diversity, the U.S. growth rates appear low compared to those of many other countries. Should we regard the U.S. performance as unsatisfactory and in need of remedial U.S. saving and investment policies? Should we expect the United States to grow as fast in per capita terms as the current leaders or, at least, to approach the average achieved by other industrial countries?

What Is a 'Fast' Rate of Growth?

The way to judge what is fast economic growth is to look at the historical record. How high a rate of growth in GDP was attained in the past? How long should the reference period be? Year-to-year changes are too short to measure sustained performance. A half century is apt to embrace too many different growth experiences; moreover, we do not have adequate data for non-overlapping half centuries. Perhaps a 10- or 25-year period is more suitable—the latter is long enough to reveal persistent tendencies but short enough to provide observations over the length of the post-World War II period.

To find historical examples of rapid growth, we have to start from 1750 or 1800. Before 1800 growth proceeded at a snail's pace; long run growth rates were under 0.5 percent per annum for total output and even lower on a per capita basis.¹ The record since then shows some spurts of GDP growth exceeding 4 percent per annum over 10-year spans, but such cases are few.²

As expected, the rates for longer periods, such as 25 or 30 years, were lower. Growth rates for a number of industrial countries are shown in Table 1-2 for successive 25- or 30-year periods spanning more than a century. We also include growth rates for 1950-84. The years after 1980 were a time of slower growth for all the countries, but the relationships among the countries for 1950-1980 were not very different from those for 1950-1984.

The first two periods of Table 1-2, 1870-1900 and 1900-1925, cover years of rapid expansion in the industrial economies. Yet we see but one instance of a total GDP rate in excess of 4 percent (the United States), and only three others (Canada, Finland and Japan) in excess of 3 percent.

Therefore, by historical standards, a long-run (25- or 30-year) total GDP growth rate in excess of 3 percent per annum may be regarded as fast, and anything over 4 percent as very fast growth. We set these standards for high growth from the record for industrial countries because historical growth rates for both aggregate and per capita GDP were so much lower in the less developed countries.

The record of per capita GDP growth, a more significant gauge than total GDP, shows that the highest rates in the two early periods were only a little over 2 percent. Thus, by historical standards, a long-run per capita income growth of 2 percent may be regarded as fast, 1.75 percent in the upper range, and 1.50 percent about par for the course.

A Golden Age of Economic Growth

By historical standards, the overall expansion of the world economy in the period from 1950 to 1984 was extraordinary. The growth of *world* GDP, at 4.45 percent per annum, substantially exceeded the growth of the *fast growers* in the quarter-centuries 1900-25 and 1925-50 and also in the thirty year period from 1870 to 1900. And the growth of world per capita GDP, at 2.53 percent per annum, was substantially higher than that of the industrial countries in earlier periods except for Germany from 1925 to 1950.

The experience since 1950 is also remarkable for the diffusion of growth among the nations of the world. For both aggregate and per capita GDP, each of the three main groups of countries shown in Table 1-1 grew at or near historically fast rates. The GDP growth of the LDCs, which was slow in the late 19th century, accelerated sharply from 1950 to 1984. Thus, development economists who once wondered whether there was something about economic develop-

---

¹See Kuznets (1971), p. 25. Maddison's estimates (1982, pp. 6-7) are still lower.

²Maddison's data (1982) in the form of annual figures extending back as far as 1820 for 16 developed countries. If the record is scanned for shorter periods, higher rates are observed. For 4- or 5-year periods, for example, instances of per annum rates of 5 percent and even 7 percent are found for Germany, Japan, and the United States within the period from 1873-1953.

Table 1-2: Historical Growth Rates for Selected Industrial Countries

	1870-1900	1900-1925	1925-1950	1950-1980	1950-1984
Per annum growth rate in GDP					
U.S.	4.26	3.46	2.73	3.35	3.25
Canada	3.08	3.32	3.85	4.61	4.25
Japan	2.45	3.12	.87	7.27	6.88
Belgium	2.02	1.49	1.06	3.50	3.12
Denmark	2.41	2.69	2.67	3.31	3.20
Finland	3.08	2.31	2.79	4.42	4.21
France	1.63	1.24	1.12	4.56	4.16
Germany	2.74	1.65	1.80	5.37	4.81
Italy	.89	2.18	1.41	4.84	4.33
Netherlands	2.03	2.66	2.15	4.24	3.70
Norway	1.86	2.55	3.21	4.13	3.83
Sweden	2.49	2.85	3.09	3.33	3.01
U.K.	2.06	.91	1.79	2.36	2.26
Per annum growth rate in GDP per capita					
U.S.	2.04	1.73	1.61	1.97	1.91
Canada	1.78	1.03	2.25	2.68	2.41
Japan	1.61	1.89	- .45	6.05	5.73
Belgium	1.09	.90	.63	3.04	2.72
Denmark	1.38	1.51	1.77	2.69	2.65
Finland	1.68	1.40	2.00	3.81	3.60
France	1.45	1.07	1.00	3.71	3.33
Germany	1.66	1.15	2.76	4.65	4.19
Italy	.22	1.42	.68	4.21	3.75
Netherlands	.83	1.19	.86	3.08	2.62
Norway	1.01	1.70	2.50	3.35	3.10
Sweden	1.79	2.17	2.48	2.77	2.50
U.K.	1.13	.54	1.34	1.99	1.94

Sources: 1870-1950: Maddison (1982).
　　　　 1950-1984: See appendix.

Per annum growth rates between terminal years.
Note: Corrections for changes in frontiers were made by Maddison to the figures for GDP, but not to those for population. Additional corrections have been made here for population for Belgium, Denmark, France, and Germany.

---

ment that prevented its spread beyond northwestern Europe, its ethnic offshoots, and possibly Japan, now see no reason to believe that economic growth is culturally bound.

This does not mean that all countries had similar growth experiences. On the contrary, within each group striking differences are found. For example, some developing countries, mainly in sub-Saharan Africa and on the South Asia subcontinent, had little per capita GDP growth or moved even further into poverty. But growth has been the more typical story in the years since World War II. Almost two thirds of the world's population live in countries in which per capita GDP grew by 2 percent per annum or more from 1950 to 1980. This includes nearly half of the people living in developing market economies. For these people, the important consequence of economic growth was not simply more goods but lower infant mortality, longer life expectancy, and sharp increases in the primary school enrollment rate. For example, between 1950 and 1982, the average life expectancy at birth increased by nine years in industrialized countries, eight years in "middle-income" developing countries, and almost 25 years in low-income countries. The adult literacy rate rose, between 1950 and 1975, by four percentage points in the first group, 23 in the second, and 16 in the third.[3]

What Growth Rates Shall We Demand of the Economy?

Granted that the experience of the last few decades in the world economy is unique in important respects, it may be asked whether it provides better guidelines for expectations about future performance than longer run economic history. For the developing countries, clearly a stagnant or very slow growing past is not relevant and more recent guidelines must be sought. But what about the United States? Should our assessment of its future growth be based on the longer sweep of history or on the higher growth rates that other industrial economies have more recently demonstrated?

The long-run average growth rate of real per capita GDP for the United States has been near 2 percent. This rate also characterized the 1950 to 1984 period, in which the average for industrial countries as a whole was near 3 percent. If the recent level of U.S. population growth, around 1 percent per year, is regarded as unlikely to change substantially, these per capita rates, if they were targets for future U.S. growth, would correspond to 3 and 4 percent in aggregate GDP growth, respectively. It might be argued that the more recent rapid growth experience of other countries is more akin to the conditions underlying current U.S. growth than the performance of the U.S. economy many decades ago. Both supply and demand factors could be drawn upon to support this view. It could be argued that the pace of innovation has quickened and the diffusion of technology has both broadened and accelerated. From the demand side, powerful pressures for faster growth have emerged that are reflected in the rise of welfare programs in affluent countries, such as the United States. The consequence is that the pursuit of high economic growth is becoming institutionalized, perhaps more in other countries than in the United States but in the United States, also. Overall, such changes may have rendered past growth rates obsolete as a basis for assessing the present or the future.

[3]World Bank (1980), Table 4.1.

The growth targets that are drawing the most attention are those that have emerged in an international context, some relating mainly to the short run. In an interdependent world, slow growth in one country or in a group of countries that constitutes a major market for others is seen as a barrier to the achievement of high growth by trading partners. This has been evident in attempts to coordinate macroeconomic policy among the leading industrial countries at summit meetings of heads of state and, between meetings, in interchanges of exhortations at lower levels. The exchange rate has emerged as the main focus of these meetings, but the desire to meet higher GDP growth targets is also voiced. As a world leader, the United States has campaigned vigorously for faster growth in its trading partners so as to ease its trade deficit and to hold in check strong domestic pressures for protection.

More explicit growth rate targets have emerged in connection with LDC debt. The idea that rapid growth in the industrial countries is important to the economic success of the LDCs is hardly new, but it has been given urgency by a succession of financial strains—oil price increases in 1973 and 1979, lagging commodity prices, high real interest rates and world recession. All of these factors contributed to the expansion of developing country debt from 14.1 percent of GNP in 1970 to 33.8 percent in 1984 (World Bank, 1985, p. 24). A major element in the ability of the developing countries to meet their debt obligations is the buoyancy of world markets for their exports, the key to which is the growth rate of the GDP of the industrial countries. Before the collapse of the OPEC cartel and the fall in interest rates in 1985-86, a widely accepted theory was that a growth rate of 3 percent for OECD countries in 1984-86, together with other plausible assumptions about exchange rates and inflation, would make the debt problem manageable. This short-term target seems modest in terms of the historical record of the developed countries.

There are somewhat more ambitious growth rates in the "high" scenario set out in a 1985 World Bank report that examined alternative prospects for the world economy. "High" and "low" scenarios for the period from 1985 to 1995 are based on the following GDP growth rates:

	Aggregate		Per Capita	
	High	Low	High	Low
Industrial countries	4.3	2.5	3.7	2.0
Developing countries	5.5	4.7	3.5	2.7

Source: World Bank, 1985, p. 138.

The high scenario involves aggregate GDP growth rates that slightly exceed those of the 1950 to 1984 period; the difference is substantial with respect to the rates of per capita GDP growth.

Whether the 1950-84 growth rates or those of the late nineteenth and early twentieth century should be taken as attainable gauges of fast growth depends on many complex interacting factors that cannot covered in detail in this report. Some of these factors will be considered in Chapter 4.

The Relative Stability of U.S. Long-term Growth Rates

With the exception of the United States, the individual countries in Table 1-2 had 1950-84 per annum growth rates for both aggregate and per capita GDP that exceeded those of any of the three quarter-centuries within the period 1870-1950, often by substantial amounts (two or more percentage points). (The only exception other than the United States was Sweden's aggregate GDP growth in 1925-50.) The experience of the United States thus represented less of a departure from its past record; its growth rate for 1950-84 was near the average for the previous three periods—higher than the immediately preceding period, nearly the same as that for 1900-25, and lower than in the 1870-1900 period. Among the twelve other countries, Sweden, and to a lesser degree, Denmark and the United Kingdom, share this U.S. pattern of near continuity in the level of growth in 1950-84 rather than a sharp acceleration. Except for the United Kingdom, the rates of GDP growth of those countries for the successive quarter-centuries between 1870 and 1950 were very respectable, averaging near 3.5 percent for the United States and over 2.5 percent for Sweden and Denmark. However, a growth rate near 3.5 percent per annum, which placed a country at the top of the list of fast growers in the earlier periods, put it in the lower half of the distribution in the rapid expansion of the three decades beginning in 1950.

These findings are even more pronounced in per capita terms. The U.S. per capita GDP growth rate for 1950-80 and 1950-84 deviated less from those of earlier periods than those of all other countries.[4] A decline of U.S. population growth tended to offset the drop in the GDP growth rate over the successive periods. In fact, the per annum growth of real per capita GDP was the same— near 2 percent—in 1950-84 as in 1870-1900, and not radically different from the rates of 1900-25 and 1925-50. But here again, a growth rate that would have given the United States a high or at least middling growth rank in an earlier period places it as one of the slowest growers in 1950-84.

A reasonable reading of the historical record is that the United States had high growth rates in aggregate and per capita GDP over long periods from 1870 to 1950 relative to other industrial countries and continued on much the same historical growth path in the 1950-84 period. This is not to suggest that growth has been steadier in the United States in all respects. Annual fluctuations of the growth in GDP around the long-period averages are relatively large, and the decline of U.S. GDP in the Great Depression of the 1930s was greater than that of other countries. But if attention is concentrated on average growth rates for long-run periods, the U.S. rate does seem high, and there is less variability in the long-period averages than for other industrial countries.

---

[4]Similar conclusions emerge if the previous century is divided into different historical periods. Using historical periods selected by Maddison, for example, his 1950-79 per capita growth rate for the United States is closer to the U.S. rate for 1820-70, 1870-1913, and 1913-50 than is the case for eight or nine other industrial countries for which similar comparisons could be made.

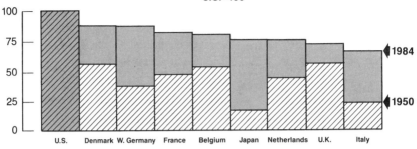

Chart 1-2
Catching Up to the United States
Real GDP per Capita
U.S.=100

Source: Table 1-3.

Beginning with the 1950s, there was a period of catch-up growth among the industrial countries. Among those in Table 1-2, only the United Kingdom had a per capita growth rate that was as low as that of the United States. Industrial countries as a whole had a per capita growth rate nearly 1 percentage point above the U.S. level. The catch-up tendency also shows up clearly in Chart 1-2 and Table 1-3, where the growth rate in real GDP per capita between 1950 and 1984 is related to the level of real GDP per capita (U.S.=100) in 1950 for the eight industrial countries for which benchmark studies were made for both 1950 and 1980.[5] It can be seen that those with lower 1950 levels made greater gains.[6]

To sum up, the U.S. economy over the period from 1950 to 1984 was like a ship moving at its familiar moderate speed, while most of the other vessels that once lagged far behind began to accelerate. Before the world enjoyed the high and widely diffused growth rates of the post-World War II era, the U.S. growth rates seemed quite favorable, but in the post-war environment other countries became the star performers. What changed was an explosion of growth in the rest of the world. What has to be explained is why per capita GDP in much of the rest of the world surged, while the U.S. rates deviated little from their historical levels.

---

[5] No benchmark estimate is available for Japan for 1950, and Japan is entered as an addendum item in Table 1-3. The inclusion of Japan greatly strengthens the catch-up story. Based on a rough approximation, Japan's real GDP per capita was around 17 percent of the U.S. level in 1950 (Kravis, Heston and Summers, 1982, p. 327); it expanded to 73.5 percent of the U.S. in 1980 according to the ICP Phase IV benchmark study (U.N. and Commission of the European Communities, 1986).

[6] When 1950-84 growth rates for industrial countries are regressed against their 1950 real GDPs per capita, the R^2 is .72. A similar regression for all countries for which data were available (115) did not reveal any significant relationship.

Table 1-3: The Catch-Up of Industrial Countries Relative to U.S., Real GDP per Capita, 1950-84

	Indexes of real GDP per capita*		Change in Index	Growth rate of real GDP per capita in constant prices of each country
	1950 (1)	1984 (2)	1984-1950 (3)=(2)-(1)	1984/50 (4)
1. Italy	23.9	66.2	+42.3	3.75
2. Germany	37.6	88.9	+51.3	4.19
3. Netherlands	45.1	75.8	+30.7	2.62
4. France	47.5	83.2	+35.7	3.33
5. Belgium	54.0	81.2	+27.2	2.72
6. Denmark	55.9	89.4	+33.5	2.65
7. U.K.	57.4	73.4	+16.0	1.94
8. U.S.	100.0	100.0	0	1.91
Addendum				
9. Japan	17.1	76.3	+59.2	5.73

*Based on PPP conversions from benchmark studies, except for Japan in 1950, which is based on PPP extrapolated from a 1975 benchmark.
Sources: Column 1; Summers, Kravis and Heston (1980), p. 30, except for Japan, which is from Kravis, Heston, and Summers (1982), p. 327. Column 2; U.N. and Commission of the European Communities (1986), extrapolated from 1980 to 1984 using real GNP per capita. Column 3; Columns 1 and 2. Column 4; Table 1-2.

Within-Period Variations in Growth

Although our interest is in long-term trends, and we have therefore treated the period since 1950 as a single unit, it is important to consider whether variation in growth patterns within the period from 1950 to 1984 would lead us to alter the conclusions we have drawn on the basis of data for the whole period.

A great deal of attention has been focused in recent years on the slowdown of growth in GDP and more particularly in the growth in productivity, since 1973.[7] This phenomenon has been discussed mainly with respect to the industrial countries. Even for them, however, growth rates in the late 1970's were not very low in a longer run historical perspective. (See Table 1-4.) From 1973 to 1980, the growth rate of the industrial countries was 2.47 percent per annum for aggregate GDP and 1.74 percent on a per capita basis. The figures in the table for the 1973 to 1984 period, though calculated like our other growth rates from the beginning and ending years of the period, include the effects of the recession in the early 1980's when U.S. real GDP actually declined in two of the years.

[7] See, for example, Lindbeck (1983), Giersch and Wolter (1983), Denison (1983), and Kendrick, ed. (1984).

Chart 1-3

The Slowdown in Economic Growth, 1950-84

Annual Percent Change in Output per Capita

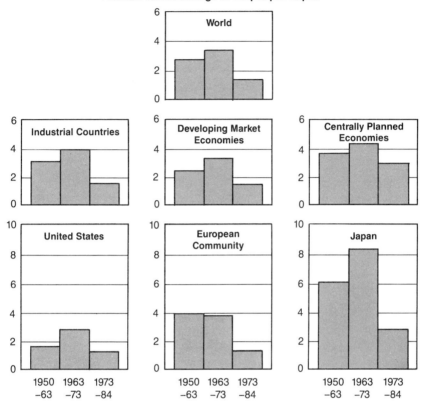

Source: Table 1-4.

For the other groups of countries, the middle and late 1970s were years of good or even excellent growth and the inclusion of the early 1980s does not alter the picture. For the period 1973 to 1984, aggregate GDP grew at rates of 4.27 percent and 4 percent for centrally planned economies and developing countries, respectively.

Within the 1950 to 1984 period, one set of growth rates is made to seem low in contrast to an adjacent set of very high rates similar to what is observed between the different quarter-centuries. Here it is the growth rates of the 1970s that appear low, particularly those of the industrial countries. The extremely high rates that produce this effect are those of the previous decade, from 1963 to 1973 (Chart 1-3). Even the slowest per capita growth rate for the industrial

Table 1-4: Annual Growth Rates of Total and Per Capita GDP, Selected Countries and Areas, and Selected Periods, 1950-84

	GDP Growth (%)			
	1950-63	1963-73	1973-80	1973-84
Industrial Countries (21)	4.34	4.96	2.47	2.32
United States	3.42	4.03	2.24	2.33
Canada	4.73	5.73	2.84	2.36
European Community (10)	4.78	4.62	2.24	1.76
France	4.79	5.48	2.84	2.25
Germany	7.64	4.63	2.33	1.72
Italy	5.93	4.89	2.79	1.98
U.K.	2.47	3.23	.92	1.13
Norway	3.71	4.29	4.68	3.56
Sweden	3.77	3.79	1.87	1.41
Japan	7.36	9.64	3.83	3.89
Developing Market Economies (86)	4.68	5.92	4.94	4.01
Low Income Countries	3.97	3.03	3.82	3.99
India	4.05	3.21	3.82	4.32
Middle Income Countries	4.97	6.85	5.23	4.01
Oil Exporters	4.91	7.45	6.18	4.19
Relatively Industrialized	4.97	6.52	5.71	4.17
Brazil	6.52	8.17	7.01	4.46
Korea	5.42	9.66	7.53	7.50
Mexico	5.68	8.01	6.43	4.53
Other	5.00	7.16	3.45	3.55
Centrally Planned Economies (8)	5.43	6.65	3.84	4.27
China	5.74	9.26	5.64	6.69
USSR	5.49	5.79	2.68	2.66
World (115)	4.65	5.57	3.39	3.22

countries entered in Table 1-4 for 1963 to 1973, 2.75 percent per annum for the United Kingdom, would produce almost a doubling of income if sustained for 25 years. Growth rates for the preceding years, from 1950 to 1963, were also very high.

In many respects, the patterns described earlier for the entire period characterized the subperiods as well. Industrial country growth rates in aggregate GDP were consistently lower in all four sub-periods than the rates of the other two major groups. Among the industrial countries shown in Table 1-4, Japan consistently had one of the the highest and the United Kingdom the lowest growth rate. Germany, whose growth rate rivaled that of Japan in the 1950s, grew at slower and slower rates. On a per capita basis, the centrally planned economies consistently had the highest growth rates followed in the first two subperiods

Table 1-4 (concluded)

	Per Capita GDP Growth(%)			
	1950-63	1963-73	1973-80	1973-84
Industrial Countries (21)	3.10	3.95	1.74	1.62
United States	1.71	2.86	1.20	1.29
Canada	2.16	4.13	1.59	1.14
European Community (10)	3.96	3.89	1.98	1.51
France	3.74	4.58	2.40	1.73
Germany	6.50	3.83	2.43	1.83
Italy	5.25	4.18	2.35	1.61
U.K.	2.00	2.75	.90	1.12
Norway	2.79	3.49	4.20	3.12
Sweden	3.17	3.09	1.57	1.19
Japan	6.08	8.44	2.69	2.90
Developing Market Economies (86)	2.35	3.36	2.43	1.52
Low Income Countries	1.91	.60	1.37	1.53
India	2.05	.85	1.49	2.01
Middle Income Countries	2.37	4.22	2.67	1.49
Oil Exporters	2.55	4.89	3.48	1.49
Relatively Industrialized	2.37	3.98	3.32	1.86
Brazil	3.33	5.39	4.54	2.09
Korea	3.13	7.15	5.83	5.82
Mexico	2.48	4.57	3.32	1.61
Other	2.20	4.29	.77	.88
Centrally Planned Economies (8)	3.83	4.38	2.47	2.99
China	4.08	6.36	4.03	5.19
USSR	3.70	4.70	1.78	1.77
World (115)	2.79	3.42	1.57	1.44

Source: See appendix.

by the industrial countries and in the period from 1973 to 1980 by the developing market economies. But the extension of the period to 1984 brings down the relative growth rate of the developing countries.

This effort to place world growth rates in historical perspective suggests that what has to be explained is the fast growth of the 1950s and 1960s rather than the slower growth rates of the 1970s that were much closer to the historical record. The United States has more nearly maintained its historical growth rates, while growth has fallen in the other countries from extraordinarily high levels. The factors that explain why there was more rapid growth abroad from 1950 to 1984 than in the U.S. may also help to account for the temporal pattern of the differences. The growth rates of the other industrial countries were farther above that of the United States from 1950 to 1973 than afterwards. These factors, which will be considered in Chapter 4, include: 1) the existence of a technological gap that was presumably larger at the beginning of the period; 2) the development of modes of transferring technology more extensively and with less delay; and 3) the impact of trade liberalization. The willingness of the United States to see its partners prosper through trade and aid was a strong influence in promoting expansion in the early post World War II years.

Up to this point, we have not reviewed or assessed the work of those who have attempted to explain the slowdown in the rate of productivity growth, particularly since 1973. Among the more recent, these include Baily (1981), Berndt (1984), Darby (1984), Denison (1985), Giersch and Wolter (1983), Jorgenson (1984), Lindbeck (1983), and Maddison (1984). The omission is, in part, because the discussion would sound much like our discussion of the decline in the growth of output per capita. We give priority to the latter because it comes closer to a welfare measure. Productivity growth is important because it can provide the means to reach higher levels of per capita income but it is not desirable in itself. The United States could increase its average productivity by greatly raising the minimum wage so that many of the least productive workers would become unemployed. Output per worker would rise, but output per capita would fall.

Nevertheless, the slowing of per capita income growth that we observe after 1973 must mainly reflect a slowing of growth in output per worker. And in looking toward the future, we would expect that judgments about future rates of productivity growth are also judgments about rates of per capita income growth. The post-World War II period was a golden age for productivity growth, just as it was for per capita income growth. The reasons we have given for considering the historical rates of growth in world per capita income a better guide to the future than the rates of 1950-80 or 1950-73 apply to productivity as well as to income.

There are two issues to be considered with respect to the record of productivity, compared to that of per capita income or output. One is whether it implies a different conclusion regarding the world's rate of growth, the exceptional characteristics of the period from 1950 to 1973, or the most likely prospects for the future. The second issue is whether the comparison between the United States and the other industrial countries looks different in productivity terms. The two questions are not completely independent because one reason offered for the world productivity slowdown is that the large productivity gap between the United States and other countries that existed just after World War II permitted very fast growth on the part of other countries as they absorbed U.S. technology, but that the narrowing of the productivity gap reduces the scope for catch-up growth.

It is difficult to summarize the productivity literature. This is partly because there are so many different concepts of productivity and partly because there is no consensus on whether or not there has been a slowdown and, also, what caused it. (Darby, for example, questions the evidence that there has been a slowdown.)

We can summarize our view of the productivity issue as follows: With respect to the growth for the world as a whole, there has been a slowdown relative to the period of most rapid growth, but the recent record does not compare unfavorably with the longer historical record. As with the growth of income per capita, we think it is the longer historical record that should guide our expectations for the future.

The comparison between the United States and other industrial countries is a more complex issue. In this case, the definition of productivity chosen makes a difference in the interpretation. The U.S. performance looks worst when crude measures, such as output per worker, are used. The difference between the United States and other countries is smaller if one measures productivity by output per unit of labor input, that is, counting the inexperienced workers (mostly female) added to the U.S. labor force in much larger numbers than in other countries, as representing less input of labor than experienced workers. And the difference is still smaller if one measures productivity by output per unit of labor and capital input. In particular, the difference between the United States and Japan is not so much a matter of faster growth of productivity in Japan as of a faster rate of substitution of physical capital for labor, as Japan invested much more heavily than the United States in new capital equipment.

To the extent that the U.S. capital stock was older on average than that of most European countries and Japan, it was probably more energy intensive and suffered more of a loss in value from the two oil shocks than did the capital stocks of the other countries. The effect of energy price increases was another form of reduction in the growth of capital relative to labor that slowed the rise in output per worker.

Conclusions

The period from 1950 to 1984 was a golden age of economic growth in the world economy, particularly the years up to 1980. The growth of world real GDP and of real GDP per capita was very rapid by historical standards. It was also widely diffused among the different countries of the world; most developing countries and planned economies enjoyed high growth rates and, at a slightly lower average level, so did industrial market economies. This is not to say that there was not a diversity of experience within each of these three major groups. Among the industrial countries, the United States was a relatively slow grower. From 1950 to 1984, it maintained a per annum total GDP growth rate of 3.25 percent and a 1.91 percent rate of growth of per capita GDP. These rates, which compare with averages of 3.86 and 2.86 percent for the industrial countries as a whole, were not radically different from those that characterized the U.S. economy in the three preceding quarter centuries. However, in those earlier periods growth rates of more than 3 percent for aggregate GDP and 2 percent for per capita GDP were exceptional and the United States was a growth leader.

The U.S. fall from the top growth ranking among the industrial countries in 1870-1900 and 1900-25, and from a middling position in 1925-50 to near the bottom after World War II, is thus attributable not to changes in the U.S. growth rates but to an explosion of economic growth in other countries. Furthermore, within the industrial group there has been a tendency since World War II for the countries that initially had lower incomes to grow faster than others. The result has been that other countries have caught up with the high per capita income level of the United States to the point where several smaller countries—even if we exclude the richly endowed oil exporting countries with small populations—enjoy per capita incomes almost as high as that of the United States.

If the historical experience of the United States is taken as the guideline, a per capita GDP growth rate of 2 percent per annum, almost doubling income levels from one 35-year generation to the next, would be regarded as very satisfactory. If we judge by the performance of other industrial economies in the period after World War II, we might raise our expectations to 3 percent. But if the rapid growth in other countries did depend on their assimilation of U.S. technology, such a rate might be out of reach. On the other hand, the spread and intensification of R & D efforts may accelerate the growth of all countries.

Appendix 1A
Some Fine Points ...
Problems in Comparing Growth Rates

Without a detailed investigation of the methods by which each country makes its income estimates and transforms them from a current price to a constant price, or real basis, we cannot say how comparable the GDP growth rates of different countries really are. Among the many possible sources of problems in making comparisons, some are particularly noteworthy:

1. There may be differences in the definition and coverage of economic activities despite the wide use of definitions provided in the U.N. *System of National Accounts*. The underground economy may differ in importance and escape measurement to varying degrees. The shift of activities to the market from the subsistence sector of developing countries and from the household sector of both developing and developed countries may produce, to varying degrees in different countries, increases in measured national output and income that may not represent net increases in welfare. This applies, for example, to services formerly performed by housewives and now commercialized such as cooked or partially cooked frozen foods. A country with a rapidly shrinking subsistence sector or with a marked influx of women into the labor force may consequently wind up with a higher reported growth rate in per capita income.

2. Changes in the output of certain sectors, notably services rendered by employees in government or in education, are typically measured by changes in the number of personnel, because there is no readily available measure of output. Countries differ in the assumptions they make about changes in the productivity of these inputs, some assuming constant productivity and others assuming some arbitrary increase per annum. Such services were typically near 10 percent of GDP in rich countries and nearly twice that proportion in poor countries in 1975.[8] As between two rich countries, with all other things equal, one that assumed a 2 percent increase in the productivity of such inputs would have a measured growth rate in national product 0.2 percent higher than a country assuming no increase.

3. The price indexes that are used to deflate current value series for producer durables, consumer durables and construction are likely to be biased upward to varying degrees from country to country. Some countries do not have special price indexes for these categories. Countries that do maintain special indexes often base them on dubious proxies, such as the prices of basic materials (e.g., steel) for producer durables price indexes or the prices of construction materials and wage rates for construction price indexes. Unfortunately for our purposes, this makes comparisons of growth rates for investment goods particularly suspect.

4. The measured growth of real output varies according to the prices used to value each year's quantities. In general, the more remote in time the selected prices are, the greater the growth rate tends to be. Today's quantities are adapted to today's price structure; if valued at an old set of prices, goods that were once expensive but are currently consumed in large quantities because they are cheap, will inflate today's constant price aggregate. A common practice is to limit these effects by altering the base year for constant prices every 10 years, but even where this is done its timing is not uniform among countries. Hence the base year that is the reference year for the weights used to construct the price and quantity indexes differs from country to country. International agencies and others compiling the data of different countries usually shift the data of all the countries to a common "base" date in a more superficial sense by dividing each constant price series by the value for the selected "base" year. This does increase comparability, but does not compensate for the fact that prices from different years were used by different countries to obtain their constant price series.

The effort to quantify the factors that hinder comparisons would entail major research. In the absence of such an effort we offer our judgment that these shortcomings are sufficiently great to discount differences in growth rates between two developed countries that are less than 0.5 percent per annum and that are less than 1 percent per annum between a developed country and a developing one. If anything, the margins of error suggested are modest.

---

[8]Kravis, Heston, and Summers (1982), p. 140.

Appendix 1B
Data Sources

Real GDP per capita estimates for 58 countries for 1980 in 1980 international dollars (I $)[9] were obtained from a U.N. tape giving the detailed results of Phase IV of the International Comparison Project (ICP), (United Nations and the Commission of the European Communities, 1986). The real GDP per capita of each country was obtained by summing the expenditures in international dollars on the 151 expenditure categories into which GDP is divided in the U.N. study. The real GDP per capita figures for the 58 countries were also used to select an estimating equation that provided estimates for 66 other market economies. The independent variables, obtained from a December 1985 World Bank national accounts tape, were exchange-rate converted per capita GDP and openness (the ratio of exports plus imports to GDP). The 1980 PPPs for these 124 countries were calculated by dividing nominal (own currency) GDP by real GDP and then multiplying by the exchange rate. These 1980 PPPs were extrapolated to 1984 on the basis of the change in the given country's GDP implicit deflator relative to the change in the U.S. implicit deflator. The 1984 PPPs were used to convert 1984 nominal GDP per capita into real GDP per capita in 1984 I$. World Bank population data were used to calculate real GDP in 1980 and 1984.

Estimates of real GDP per capita for centrally planned economies were obtained by extrapolating the 1975 estimates from Kravis and Lipsey (1984) to 1980 and 1984. For Bulgaria, Czechoslovakia, East Germany, Hungary, Poland and Romania, data from Alton et al (1985) were used for extrapolation. For the U.S.S.R. the extrapolations were based on C.I.A. estimates and for China on "national income" as reported in *The Statistical Yearbook of China, 1984*. The 1975 U.S. estimate was extrapolated using World Bank data, and the ratio of each country's real GDP per capita relative to the U.S. was calculated. To convert these ratios to international dollars, they were multiplied by the U.S. per capita GDP for 1984.

Population data in 1980 and 1984 were obtained from the World Bank tape for all countries except China, for which National Institutes of Health data were used.

These procedures yielded 1984 estimates on real GDP per capita for 132 countries which were classified as follows:

---

[9] An international dollar has the same purchasing power over GDP as a whole as a U.S. dollar. Its purchasing power over components of GDP is not necessarily the same as that of a U.S. dollar because relative prices of components of GDP reflect world average price relationships among components.

a. Industrial countries (as listed in IMF *International Financial Statistics,* June, 1984, p. 18) (21)—Australia, Austria, Belgium, Canada, Denmark, Finland, France, Germany, Iceland, Ireland, Italy, Japan, Luxembourg, Netherlands, New Zealand, Norway, Spain, Sweden, Switzerland, United Kingdom and United States.
b. Centrally planned economies (8)—Bulgaria, China, Czechoslovakia, East Germany, Hungary, Poland, Romania and U.S.S.R.
c. Oil exporters (as listed in IMF *International Financial Statistics,* June, 1984, p. 18) (9)—Algeria, Indonesia, Kuwait, Libya, Nigeria, Oman, Saudi Arabia, United Arab Emirates and Venezuela.
d. Relatively industrialized countries—countries for which World Bank data indicated that manufacturing comprised at least 20% of total GDP in 1980 (17)—Argentina, Brazil, Chile, Egypt, Hong Kong, Korea, Malaysia, Malta, Mexico, Nicaragua, Peru, Philippines, Portugal, South Africa, Turkey, Uruguay and Yugoslavia.
e. Low income countries—countries with 1980 real GDP per capita less than $1000 (32)—Bangladesh, Benin, Burkina Faso, Burma, Burundi, Cameroon, Cape Verde, Central African Republic, Ethiopia, Gambia, Guinea-Bissau, Guinea, Haiti, India, Kenya, Madagascar, Malawi, Mali, Nepal, Niger, Rwanda, Senegal, Sierra Leone, Somalia, Sudan, Tanzania, Togo, Uganda, Yemen, Zaire, Zambia and Zimbabwe.
f. Other middle income countries (45)—Antigua and Barbuda, Bahamas, Bahrain, Barbados, Belize, Bolivia, Botswana, Colombia, Congo, Costa Rica, Cyprus, Dominican Republic, Dominica, Ecuador, El Salvador, Fiji, Ghana, Greece, Grenada, Guatemala, Guyana, Honduras, Iran, Israel, Ivory Coast, Jamaica, Jordan, Liberia, Mauritania, Mauritius, Morocco, Pakistan, Panama, Papua New Guinea, Paraguay, Sao Tome & Principe, Sri Lanka, St. Kitts-Nevis, St. Lucia, St. Vincent, Suriname, Syria, Thailand, Tunisia and Western Samoa.
g. European Economic Community (10)—Belgium, Denmark, France, Germany, Greece, Ireland, Italy, Luxembourg, Netherlands and United Kingdom.
h. Middle Income countries (71)—Oil exporters plus relatively industrialized plus other middle income countries.
i. Developing market economies (103)—Low income and middle income countries.

Estimates of real GDP per capita could not be derived by the methods described above for 45 countries because of missing data on income or openness. Therefore, to obtain estimates of world GDP in 1980 and 1984, we assumed that the per capita GDPs of four of these countries were, on average, equal to those of the low income countries and that the per capitas of the other 41 were equal to the average of the "other" middle income countries. World Bank population data were used to calculate real GDP, and the totals were added to the developing market economies category.

Appendix Table 1-1: World Population and Production, Selected Countries and Groups of Countries[a] 1984

	Population		Real GDP (I$)			
			Aggregate		Per Capita	
		%	$ bil.	%	$	US = 100
	(1)	(2)	(3)	(4)	(5)	(6)
Industrial Countries (21)[b]	734.21	15.7	9,480	47.2	12,912	84.5
U.S.	236.96	5.1	3,619	18.0	15,275	100.0
Canada	25.18	.5	376	1.9	14,938	97.8
European Community (10)[c]	272.82	5.8	3,183	15.9	11,667	76.4
France	55.09	1.2	700	3.5	12,710	83.2
Germany	61.20	1.3	831	4.1	13,580	88.9
Italy	57.03	1.2	577	2.9	10,113	66.2
U.K.	56.33	1.2	631	3.1	11,206	73.4
Norway	4.15	.1	61	.3	14,809	96.9
Sweden	8.34	.2	118	.6	14,152	92.6
Japan	120.07	2.6	1,399	7.0	11,652	76.3
Developing Market Economies (148)	2,531.34	54.1	4,959	24.7	1,959	12.8
Brazil	132.58	2.8	548	2.7	4,134	27.1
India	749.88	16.0	620	3.1	827	5.4
Korea	40.58	.9	164	.8	4,051	26.5
Mexico	76.95	1.6	331	1.6	4,307	28.2
Centrally Planned Economies (8)[d]	1,417.67	30.3	5,628	28.0	3,970	26.0
China	1,031.28	22.0	2,720	13.6	2,637	17.3
Soviet Union	275.03	5.9	2,097	10.4	7,625	49.9
World (177)	4,683.22	100.0	20,067	100.0	4,285	28.1

[a] The number of countries in each group is given in parentheses.
[b] Australia, Austria, Belgium, Canada, Denmark, Finland, France, Germany, Iceland, Ireland, Italy, Japan, Luxembourg, Netherlands, New Zealand, Norway, Spain, Sweden, Switzerland, United Kingdom, and the United States.
[c] Belgium, Denmark, France, Germany, Greece, Ireland, Italy, Luxembourg, Netherlands and the United Kingdom. However, Greece is *not* included among the industrial countries.
[d] Bulgaria, Czechoslovakia, German Democratic Republic, Hungary, People's Republic of China, Poland, Romania, and the U.S.S.R.

For the calculations of growth rates for groups of countries shown in Table 1-1, adequate time series were available for only 115 countries. The 1980 international dollar GDP of each of these countries was extrapolated backward to 1950 and forward to 1984 using World Bank data, and when necessary, data from Kravis and Lipsey (1984) or from the International Monetary Fund (IMF) *International Financial Statistics, 1985*. Population data were taken from the World Bank tape for all countries except China, for which National Institutes of Health estimates were used. For each group of countries an aggregate GDP in 1984 international dollars was summed, and growth rates between the years shown in Table 1-4 were calculated.

Chapter 2
Saving and Capital Formation in the United States and Other Industrial Countries

Is the United States a Spendthrift Nation?

Finding a Yardstick to Measure Saving and Investment

Ever since World War II, the United States has been thought of as a spendthrift nation—a country that consumes almost all its income and that saves and adds to its capital stock at a low rate, not quite the lowest perhaps but very low by world standards. This belief has been the cause of much concern in the United States, because it is thought that low saving rates have resulted in low rates of economic growth and will continue to do so. Questions have been raised also about the quality of U.S. investment. Americans are thought to have invested relatively large amounts in housing and consumer durables, which are supposedly less productive than investments in industry or construction.

If the standard data on the proportion of output devoted to gross capital formation are taken at face value, the story of U.S. investment is similar to that of income growth rates told in the last chapter. The U.S. ratio of gross capital formation to GNP since 1960 has not differed much from that of the period 1860-1938; it was 19.0 percent in the earlier period and 17.9 percent in the later one.

	Gross Fixed Capital Formation as % of GNP	
	Pre-WWII	1960-84
U.S	19.0	17.9
Australia	16.3	25.4
Canada	19.9	22.4
Japan	12.9	31.9
Denmark	11.7	22.2
Germany	17.9	23.1
Italy	12.6	20.5
Norway	14.0	29.4
Sweden	12.7	22.0
U.K.	8.6	18.2

Source: Lipsey and Kravis (1987), Table 1, derived from Kuznets (1966) and OECD National Accounts.

Saving, Investment, and Capital Formation

Many different definitions of saving, investment, and capital formation are used in discussions of this subject. In general, saving is the act of putting some part of current income aside for the future instead of using it for current consumption. Capital formation, or investment, is the acquisition of goods or services that become part of the capital stock, to be used in future production. Thus, the motives and mechanisms for saving are different from those for capital formation or investment. In some national income and product accounts, such as those of the United States, saving and investment are defined as equal for the nation as a whole. In the OECD accounts that we use for international comparisons, saving is *domestic* saving and capital formation is *domestic* capital formation, and any difference between total saving and total capital formation or investment must be met by lending to foreign countries (net foreign investment) or borrowing from them. In the OECD accounts, the sum of domestic investment plus net foreign investment is defined to be equal to domestic saving.

By and large, trends in saving rates and capital formation rates are similar for most countries over most periods. Furthermore, the ranking of countries by saving rates is very similar to their ranking by investment rates. At times, therefore, we will draw inferences about one from the other but will point out some differences between the saving and capital formation measures.

We distinguish between what we call conventional and broad concepts of saving and capital formation. The conventional definition of capital formation, as in the OECD accounts, includes purchases (or construction) of non-military structures by business firms, governments, and households, and purchases of non-military durable equipment by businesses and governments. It excludes government purchases of durable military equipment and structures, purchases of durable equipment by households or consumer durables, research and development expenditures, and expenditures for education and training. These outlays are all excluded even though they clearly fall within the category of long-lasting assets. All of them are treated as expenditures for current consumption in the conventional national accounts.

For the 40 years or so before 1910, and during the period between the two World Wars, the United States ranked very high in the proportion of output devoted to capital formation in the conventional sense. Of the 10 developed countries listed, the United States had the highest investment rate (gross domestic capital formation relative to GNP) in the 1870s and 1880s, the second highest in the 1890s and 1900s, and the third highest in 1909-1929. By the 1950s, however, the United States had next to the lowest rate, ahead only of the United Kingdom. Later decades saw the United States still at the bottom of this list, in a tie with the United Kingdom. At the other end of the distribution, Japan invested the highest proportion of its output.

The United States fell from being a leader in saving to being a laggard because the ratio of gross capital formation to total output changed little in the

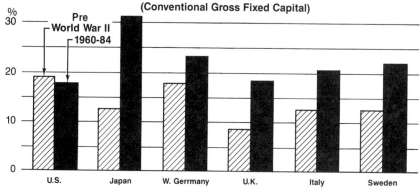

Chart 2-1
Capital Formation as Percent of GNP
(Conventional Gross Fixed Capital)

Source: Lipsey and Kravis (1987), Table 1.
Note: Periods called "Pre World War II" vary, but typically start in the 1860s and end in the 1930s.

United States between the pre-WW II years and the later period, while it increased in every other country, including even the United Kingdom, as can be seen in Chart 2-1. In the decades following WW II, even relatively low-saving countries have invested more of their incomes in fixed capital than formerly high-saving countries such as Canada and the United States had invested in the past.

The usual picture of the recent standing of the United States is based on the personal saving rate: saving by individuals as a percent of disposable personal income. This is the saving rate regularly reported by the Department of Commerce and widely publicized. Over the decade 1970-1980 the U.S. personal or household saving rate was less than two thirds of the average in eight other industrial countries for which roughly comparable data could be obtained (Table 2-1). As can be seen from the table, it makes a considerable difference whether we look at national saving or household saving and whether we look at gross or net saving.

All measures here show the U.S. saving rate to have been low in the 1970s compared to that of other countries, particularly with respect to net saving. Among 24 OECD countries the United States was the lowest on net (48% of the average of the others) and gross (76%) national saving. Gross and net household saving data are available for fewer countries. As compared with 8 other countries for which household saving rates were available, the United States national net saving rate was 45% below the average while the gross was only 24% below. The differences were smaller for the household saving rate, 37% on net, 20% on gross.

27

Gross and Net Saving and Investment

Of the total amount of a nation's output that is not used for current consumption, some part goes to replace the capital used up in production and is referred to in national income accounts as capital consumption. It is deducted from gross income or output in calculating net income or output and from gross saving and capital formation in calculating net saving and capital formation. Net investment, if correctly calculated, would measure additions to capital stock, whereas gross investment includes purchases that replace capital used up in production.

We have followed Kuznets and other writers in using gross rather than net saving and capital formation. There are several reasons for that choice. One is skepticism regarding available measures of capital consumption and, in particular their comparability among countries (Blades and Sturm, 1982). Another is the belief that the introduction of new capital equipment brings new technology into the production process, whether or not the new equipment is nominally a replacement for old equipment embodying old technology. If technology, or the quality rather than the "quantity" of capital equipment, is what drives economic growth, it is the gross rather than the net capital formation that is relevant. A country in which gross capital formation is equal to calculated depreciation, and therefore resulted in no net capital formation, would nevertheless achieve economic growth by the substitution of new for old technology.

Whose Saving Should be Measured?

This brief excursion into historical and current comparisons makes it clear that conclusions about saving ratios can be affected by both the choice of economic sectors for which saving is measured and by the kinds of provisions for the future that are counted as saving or investment. We turn first to the question of what sector's saving should be studied.

We focus on aggregate private or national saving, rather than household saving, in our international comparisons for several reasons. One is that the household data are less accurate and raise more problems of international comparability. Two aspects of household saving estimates underscore their deficiencies, even in the United States, which has relatively good statistical data. One is that the National Income and Product Accounts saving estimates often diverge by large amounts from the flow-of-funds (FOF) estimates of the Federal Reserve Board, even after adjustment for differences in definition, and trends in the two saving rates have differed substantially. The second is that revisions in the estimates of personal income have a substantial impact on estimates of saving and the saving ratio.

A source of uncertainty in the estimates of household or personal saving, even as defined in the national income accounts, is the underreporting of income on tax returns. The problem is greatly compounded by the fact that personal saving is a small residual obtained by subtracting one large amount

Table 2-1: International Comparison of Saving Rates, 1970-80 (Published Estimates Without Adjustments)

	National		Household	
	Net^a	Gross^b	Net^a	Gross^b
Saving Ratios:				
U.S., NIPA^c	6.7	16.0	7.2	12.1
SNA^d	7.8	18.6	7.7	12.6
Average of 23 other OECD countries, SNA	16.1	24.4		
Average of 8 other OECD countries, SNA	14.4	24.4	12.2	15.8
Australia	18.1	22.3	14.1	16.8
Canada	12.3	22.1	9.5	13.8
Japan	25.1	35.0	20.7	25.0
Finland	13.1	25.9	6.3	11.9
France	15.0	24.1	13.4	16.7
Italy	11.6	22.3	21.5	25.1
Sweden	12.4	21.5	4.2	8.3
United Kingdom	8.9	19.0	8.0	10.9
U.S. (SNA) as % of average of:				
23 other OECD countries	48	76		
8 other OECD countries^e	55	76	63	80

^{a}Net saving as percent of net disposable income
^{b}Gross saving as percent of gross disposable income
^{c}U.S. National Income and Product Accounts (NIPA). The main differences from the SNA concept are the treatment of all government purchases as consumption in the NIPA, whereas the SNA treats government construction and equipment purchases (except military) as investment and, in the household accounts, the treatment of estate and gift taxes as current outlays in the NIPA and capital outlays in the SNA. See Blades and Sturm (1982), p. 2.
^{d}UN System of National Accounts
^{e}These are the countries covered in our analysis for which the source shows all four saving measures

Source: Blades and Sturm (1982), p. 6.

(personal outlays) from another (disposable personal income). A relatively small error in either outlays or income can cause a very large error in a saving estimate.

In addition to these questions of accuracy and relevance, another reason to measure saving on a national basis is that saving decisions by households, business and government may be interdependent. Households may consider social security accumulation and business saving or dissaving as their own saving, and therefore take account of it or offset it by their own saving decisions. Finally, for explaining economic growth, our interest is more in aggregate or national saving and investment than in that of any single sector.

What Should be Counted as Saving?

Aside from the issue of whether to measure saving gross or net of capital consumption that we touched on earlier, the major issue in measurement is the distinction between current expenditures and capital expenditures. What kind of expenditures yield future income? And how would the inclusion of several important categories, ignored in current official saving ratios, affect the comparison between the United States and other countries?

Broad measures of investment, and correspondingly broad output measures, have been calculated for the United States by several authors.[1] Unfortunately, the broadest measures of saving and capital formation have not been estimated on a comparable basis for a large number of countries over long periods of time. In order to assess the effects of the breadth of the definition on international comparisons, we experimented with the saving ratios for the 1970s and 1980s, calculating the effects of extending the scope of the saving measure to include provisions for the future that fall outside the conventional national accounting concept of saving. Our efforts here are confined to those for which some calculations were available, or could be made fairly readily, for a considerable number of countries. We draw on the work of Blades and Sturm (1982) and Blades (1983), who had a similar objective and also sought to improve comparability among countries, but we have made our own calculations, the technical details of which are omitted here but can be found in Lipsey and Kravis (1987).

Consumer Durables

The logic of treating households' purchases of consumer durables as capital formation, corresponding to business purchases of durable producers goods, is similar to that for the treatment of house purchases as capital formation in the present systems of national accounts. Consumer durables produce services over a long period of time and, in many cases, the services are very similar to those yielded by the durables bought by business. Cars, the largest item in consumer durables, give transportation service whether they are owned by businesses or by households. Some of those owned by businesses are leased to households for their own use. Refrigerators, freezers, or laundry machines often provide services to households even if they are owned by businesses. In fact, the distinction between consumer and producer durables in the national accounts rests on ownership rather than on function.

To treat purchases of consumer durables as capital formation in the same way as purchases of owner-occupied housing, it is necessary to make two adjustments. One is to add household expenditures on consumer durables (treated as consumption in both the SNA and the U.S. national income accounts) to conventional gross fixed capital formation. The second is to add a measure of the current services yielded by consumer durables to consumption and out-

---

[1] Some alternative calculations of household saving and different views of its correct measurement can be found in Auerbach (1985), Eisner (1985), Hendershott and Peek (1985), Jorgenson and Pachon (1983), and Kendrick (1976).

put. This adjustment requires information on the stock of consumer durables and calculations of the services derived from them by consumers. Such calculations are available in detail only for the United States, but we have made some rough estimates, of little consequence numerically, for other countries.

The effect of adding purchases of consumer durables to the conventional measures of capital formation is as follows:

	Percent of GDP, 1970-84	
	Conventional Capital Formation	Capital Formation Including Consumer Durables
United States	18.1	23.1
Average of 11 other countries[a]	23.5	27.6
United States as % of average of 11 countries	77	83

[a]Canada, Japan, Austria, Denmark, Finland, France, Italy, Netherlands, Norway, Sweden, United Kingdom
Source: Lipsey and Kravis (1987), Appendix Tables 1 and 2.

This adjustment alone eliminates more than a quarter of the difference between the United States and the other countries. It also compresses the range of saving ratios slightly, since the addition is smallest for Japan and relatively large for the United States. What this suggests is that there is some substitution between investment in consumer durables and other forms of saving.

Education

Ideally, we would wish to add to conventional capital formation measures of all forms of human capital investment. These would include the earnings foregone by students while they are in school and the costs of on-the-job training. However, such data are not available in any internationally comparable form. The one part of such investment for which we can make calculations for a fair number of countries, and even these require some bold estimating, is expenditures on education. They yield a return over a long period of time in the form of higher earnings in the labor force, better care of children, and greater efficiency in consumption even after retirement from the labor force, and from child rearing. Some part of the cost of education might more properly be treated as consumption, but any overestimate of capital formation on that account should be more than balanced by our omission of foregone earnings. The effect of making this adjustment alone can be seen in the following:

	Percent of GDP, 1970-84	
	Conventional Capital Formation	Capital Formation, including Expenditure on Education
United States	18.1	24.2
Average of 14 other countries[a]	23.3	28.6
United States as percent of average	78	85

[a]Same countries as for previous table, plus Australia, Belgium and Germany.
Source: Lipsey and Kravis (1987), Appendix Tables 1 and 3.

Once again, the effect of the broadening of the concept of capital formation is to move the United States somewhat closer to the average of the other developed countries, reducing the gap by more than a quarter.

We can combine the adjustments for consumer durables and education for only 11 of the 14 countries, because data on consumer durables were not available for Australia, Belgium, and Germany. The results are only slightly affected by this limitation, since the average saving rate for the 11 countries, including education expenditures, is only slightly higher than that for the 14. The effect of the combination of the two adjustments is as follows:

	Percent of GDP, 1970-84	
	Conventional Capital Formation	Capital Formation, including Consumer Durables and Education
United States	18.1	28.4
Average of 11 other countries	23.5	32.8
United States as percent of average	77	87

Source: Lipsey and Kravis (1987), Appendix Tables 1 and 3.

Almost half of the apparent gap in investment rates between the United States and the 11 other countries is eliminated when we broaden the concept of investment to encompass expenditures on both consumer durables and education.

Research and Development

Research and development is more forward-looking than much of investment in equipment and probably includes less current consumption than education. It is thus an even stronger candidate for treatment as capital formation.

Incorporating R&D expenditures into the measure of capital formation involves in some cases an addition to the measure of total output as well, as described in the source to the table below. The effect of adding R&D expenditures, along with education expenditures, to gross capital formation is as follows:

	Percent of GDP, 1970-84	
	Conventional Capital Formation	Capital Formation Including Expenditure on Education and R&D
United States	18.1	26.2
Average of 12 other countries	23.0	29.9
United States as percent of average of 12	79	88

Source: Lipsey and Kravis (1987), Appendix Tables 1 and 4.

The broadening of the capital formation concept, as before, substantially reduces the gap between the United States and the other developed countries, in this case by over 40 percent. If we add to these two adjustments the inclusion of consumer durables expenditure, as we can do for 10 of the 12 countries other than the United States, the gap is reduced a bit further:

	Percent of GDP, 1970-84	
	Conventional Capital Formation	Capital Formation, including Expenditure on Education, R&D and Consumer Durables
United States	18.1	30.1
Average of 10 other countries	23.3	33.9
United States as percent of average of 10	78	89

Source: Lipsey and Kravis (1987), Appendix Tables 1 and 4.

Half of the gap is eliminated, and the U.S. capital formation rate is only about 10 percent below that of the other countries.

Military Capital Formation

In both the UN System of National Accounts and the U.S. National Income and Product Accounts, expenditures on defense construction and equipment are treated as current public consumption rather than as capital formation. Yet, whatever their other faults and virtues and whatever their effects on the growth of non-military output, these expenditures are intended to yield output over a long period of time. Surely, if we are interested in the extent to which a country sacrifices present consumption for future gains, these expenditures are as relevant as those for civilian capital formation.

The inclusion of military capital formation, as we would expect, raises the U.S. saving rate relative to all but one of the other countries for which we can make the comparison, and particularly with relation to Japan.

	Percent of GDP, 1970-84	
	Conventional Capital Formation	Capital Formation, including Expenditures on Education, R&D, Consumer Durables, and Military Construction and Equipment
United States	18.1	31.4
Average of 10 other countries	23.3	34.5
United States as percent of average of 10	78	92

Source: Lipsey and Kravis (1987), Appendix Tables 1 and 5.

With military capital formation included, the gap between the United States and the other developed countries is reduced by almost two thirds.

Other Adjustments to the Investment Concept

The adjustments we have made bring the U.S. investment ratio from less than 80 to over 90 percent of those of the other major industrialized countries. There are some further adjustments we did not attempt. One would be to add investment in the form of the foregone earnings of students. A recent estimate puts this item at over 60 percent of expenditures on education (Johnson, 1985). As the proportion of working-age students attending institutions of higher education is higher in the United States than in all or most of the other countries, the inclusion of this form of investment would raise the U.S. investment rate and bring it closer to the average.

Among other possible adjustments, the inclusion of more of the household economy in the accounts (in addition to owner-occupied housing and consumer durables) would probably raise the denominator (GNP) most for countries with low female labor force participation rates and thus lower their saving rates relative to countries such as Sweden. The addition of rearing costs to investment, as in Kendrick (1976), would raise saving rates for countries with relatively rapid population growth as, e.g., Canada and the United States, compared with the European countries.

As mentioned above, we have followed, with modifications, some of the procedures of the earlier comparisons in Blades and Sturm (1982) and Blades (1983). However, we have been a little freer, or perhaps somewhat reckless, in estimating missing data. In general, our results confirm theirs despite the differences in method. Both suggest that readily feasible adjustments eliminate about half or more of the observed differences in total national gross investment shares of total output between the United States and other developed countries. That leaves the United States, as measured, only about 10 percent below the average of the other countries, although still low in the ranking: ninth out of eleven in our calculations. And both suggest that the unmeasured items might well erase much of that remaining gap. Similarly, a more recent comparison of household saving rates in the United States with those of 5 other developed countries in 1980-82 by the Deutsche Bundesbank (1984) suggested that about half of the difference between the United States and the other five countries, including Japan, and more than half of the difference between the United States and the European countries, were the consequence of differences in the methods used to calculate saving.

One can think of shares of saving or capital formation in GDP as measuring some type of investment "effort" or willingness to sacrifice present consumption for future benefits. By a standard that includes both conventional and some nonconventional elements of capital formation, the United States ranks in the second half of the distribution of industrial countries and is about average within that second half. There is no evidence in these data that Americans have been substantially more "present-minded," or neglectful of future needs, than the citizens of most other developed countries.

	Expanded Capital Formation as Percent of GDP, 1970-84: Difference Between United States and Average, Excluding United States, of
10 countries	−9%
8 countries, excluding Japan and Norway	−5%
7 countries, excluding Japan, Norway, and Canada	−3%
5 lower saving countries	0%

Real Investment Ratios: How Much Capital Goods Does the Saving of Different Countries Buy?

While the share of total output in current prices may reflect a country's willingness to sacrifice present consumption to increase welfare later, it does not necessarily indicate how much capital is being acquired through that sacrifice. The reason is that capital goods are more expensive relative to other goods in some countries than in others. In particular, capital goods are relatively cheap in the United States.

Consider two countries with the same nominal saving or capital formation ratios and equal GDPs, measured in the usual way in each country's own prices. The one with the low capital goods prices will be setting aside more current output in physical terms for the production of future income, and its real capital formation ratio will be higher. Thus, it is to be expected that the U.S. real capital formation ratio will compare more favorably with that of other countries than will its nominal or own-price ratio.

The impact of these differences in price relationships on comparative capital formation ratios is summarized below for 1975 and 1980, years for which detailed comparisons of prices and real product are available for a substantial number of our countries.

Capital Formation as Percent of GDP, 1975 and 1980

	Nominal Ratios[a]		Real Ratios[b]		
	Conventional Definition	Conventional Definition	Broad Definition, including expenditure on:		
			Education and R&D	Education, R&D and Consumer Durables	Education, R&D Cons. Durables and Military Capital Formation
	(1)	(2)	(3)	(4)	(5)
			1975		
United States	16.3	16.3	24.8	29.1	30.4
Average of 8 other countries[c]	22.7	21.5	28.5	32.8	33.4
United States as percent of average	71.8	75.8	86.7	88.7	91.1
			1980		
United States	18.5	18.5	26.6	30.6	31.7
Average of 8 other countries	22.0	20.5	27.7	31.4	32.1
United States as percent of average	84.1	90.2	96.1	97.2	98.9

[a] Ratios based on own-currency expenditures on GDP and its components.
[b] World prices used to obtain the real ratios.
[c] Japan, Belgium, Denmark, France, Germany, Italy, Netherlands, and United Kingdom.
Source: Lipsey and Kravis (1987), Appendix Table 6.

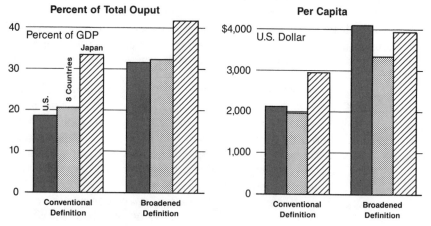

Chart 2-2
Capital Formation, 1980
(Real Gross Fixed Capital)

Source: Lipsey and Kravis (1987), Appendix Table 6.

Note: "Broadened Definition" of capital formation includes education, R&D, consumer durables and military capital. The eight countries include Belgium, Denmark, France, Italy, Japan, Netherlands, United Kingdom and West Germany.

The shift to real investment ratios raises the U.S. ratio by 4 percentage points in 1975 and 6 percentage points in 1980 (column 2). When the concept of investment is broadened to include the other provisions for future flows of goods and services described earlier, still retaining the real investment ratio concept, the U.S. ratio of capital formation to GDP rises to 91 percent in 1975 and 99 percent in 1980 of the average of the other eight countries. The inclusion of military capital formation, despite the attention it receives in public discussion, has little effect on the ratios.

Among the individual countries that make up the eight, Japan's real ratio on our most inclusive basis was still the highest, 40 percent in 1975 and 42 percent in 1980 (Chart 2-2), compared to 31 to 34 percent for Belgium, France, Germany, and the Netherlands in both years.

Since this comparison is confined to two years and to eight countries, two other points should be noted. One is that 1975 was a particularly unfavorable year for the United States. For the other countries, taken as a group, the capital formation ratio in 1975 was close to the average of 1970-84, but in the United States it was 8 percent below the period average. The other is that the comparison with these eight countries shows the United States in a less favorable light than the earlier comparisons with 11 or 14 countries. These facts suggest that a comparison of broadly defined capital formation ratios for the whole period and for more countries would almost certainly find the United States close to the average of the other countries.

Per Capita Capital Formation

When real ratios of capital formation to GDP are compared using a broadened definition of capital formation, the United States climbs into only the lower ranks of the industrial countries in 1975 and to the middle ones in 1980. That suggests that the U.S. ratio, applied to a higher income than that of most other countries, may entail more real capital formation per capita than in other countries.

If we compare the United States with the same eight countries on the basis of both conventional and broader concepts of capital formation, the results are as follows:

	Capital Formation per Capita, $U.S. at U.S. Prices, 1975 and 1980			
	Conventional	Including Expenditure on:		
		Education and R&D	Education, R&D and Consumer Durables	Education, R&D Consumer Durables, Military Capital Formation
		1975		
United States	1,172	1,805	2,343	2,448
Average of 8 other countries	1,134	1,515	1,893	1,926
United States as percent of average	103	119	124	127
		1980		
United States	2,114	3,096	3,923	4,069
Average of 8 other countries	1,978	2,701	3,283	3,347
United States as percent of average	107	115	120	122

Source: Lipsey and Kravis (1987), Appendix Table 6.

Real capital formation per capita was higher in the United States than, on average, in the other eight countries (Chart 2-2). Of the adjustments we make, the inclusion of military capital formation is the least important. Measured on the conventional basis, the U.S. level was exceeded by those of Japan, Germany, and France in both 1975 and 1980, but on the most inclusive basis, per capita capital formation was highest in the United States in both years, a little above those of Germany, the second country, and Japan in 1980 and about 15 to 17 percent above them in 1975.

For the conventional measure of capital formation, some data are available for many more countries, but at European rather than worldwide prices. If we compare per capita conventional investment in 1980 for the same 7 countries as in 1975 and for a broader group of 17 OECD countries, we find the following:

	Conventionally Defined Capital Formation per Capita at OECD Prices, 1980	
	8 Countries	17 Countries
	(U.S.=100)	
United States	100.0	100.0
Average of other countries	84.8	83.0
United States as percent of average	118	120

Source: Lipsey and Kravis (1987), Appendix Table 7.

The U.S. margin over the other eight countries in 1980 was higher than in 1975 and exceeded the average of 17 OECD countries by a little more, partly because some quite low income countries were added to the list. Conventionally defined real capital formation per capita in the United States was surpassed only in Japan (by 14 percent) and in Germany (by 7 percent) among the nine countries in these calculations. In the longer list of countries, Canada and Norway had the highest per capita capital formation, over 25 percent and about 14 percent above the U.S. level, respectively.

On the whole, the evidence points to the conclusion that the United States is close to the average of other developed countries in the degree to which it has used its income for forward-looking purposes—capital formation in a broad sense. Since the price of capital goods has been relatively low in the United States, the ratio of real capital formation, broadly defined, to real output has compared more favorably with that of the other countries than investment ratios based on each country's own relative prices. And since the United States is a high-income country, the same fraction of output devoted to investment as in other countries has kept the United States investing more per capita than the others on the average, and more per capita than almost all the individual countries.

Has Low Saving Made the United States Poor in Conventional Capital?

We can think of the purpose of saving and investment as being the accumulation of the productive wealth that leads in turn to a high level of production and consumption. A country's wealth at any time incorporates the results of saving over many years. It thus provides a summary of saving history, although wealth can be acquired in other ways, such as by changes in the value of assets already acquired.

Unfortunately, wealth figures are imperfect, as are saving data. One reason is that calculations of wealth are almost always many years out of date. Furthermore, records of wealth are kept in each country's currency and must be translated into a common currency for comparison. We do have very complete records of the rate at which at least the major currencies exchange in world markets, but, as has been pointed out, translation by this method does not produce comparability among countries.

If net saving in the United States, conventionally defined, has been low since World War II relative to that of other developed countries, we might wonder whether the United States would by now find conventionally defined capital relatively scarce. Has the United States lost its position as the most capital-abundant country?

The largest and most carefully assembled collection of asset (and liability) data from national sources is that of Raymond Goldsmith (1985). From these and a set of OECD estimates, we can make a comparison of capital per person among a dozen countries in the late 1970s and 1980. Using the conventional measure of net capital stock, we find that the United States, despite two or three decades of low conventional saving rates, still had a high level of conventionally defined capital per person in the late 1970's and in 1980. Although exceeded by a few small countries, it was the highest among the major countries, with the gap remaining very large between the United States and the United Kingdom, France, Japan, and Italy. Where both net and gross capital stocks were available from the same source, the gap was larger for gross stocks than for net. It may be that the U.S. capital stock is older than most others, or that it depreciates faster, or that a faster rate of depreciation is used in the U.S. calculations.

On the grounds mentioned earlier, we are inclined to emphasize the broadest measures of capital we can find (and regret the narrowness even of these) with respect to capital formation. We therefore prefer measures that include at least consumer durables, on the belief that ownership of a car, for example, by a household does not mean that the car performs services substantially different from those provided by taxis or public transportation, although the purchase of a car is not counted as investment and its services are not included in income and output in official calculations.

It is the high level of consumer capital that differentiates the United States most from other countries (Chart 2-3).

Chart 2-3
Stock of Consumer Durables
Per Capita, Late 1970's, U.S.=100

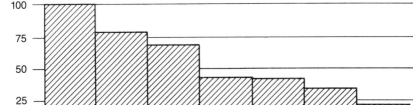

Source: Lipsey and Kravis (1987), Appendix Table 9.

	Real Net Stock of Consumer Durables, per Capita, Net, Late 1970s (U.S.=100)
United States	100
Canada	79
Japan	21
Belgium	60
Denmark	51
France	34
Germany	68
Italy	43
Norway	32
United Kingdom	42

Source: Lipsey and Kravis (1987), Appendix Table 9.

When those assets are added to the conventional capital stock, the United States is shown to have had more capital per person than any country except Canada (about equal to the United States), and to have led the major industrialized countries by 20 percent or more.

Although most of these measures show that the major foreign countries had not yet caught up with the U.S. endowment of capital by the mid- or late 1970s, consistent measures over the 1950 to 1976 period show every country gaining on the United States in conventional nonresidential capital per person (Chart 2-4).

Unfortunately, we have little data on stocks of capital outside of conventional capital and consumer durables. But we can get some indication of the stock of educational capital from data on years of education at various levels. The high level of U.S. capital formation in this form is reflected in the higher average years of formal education of the population, as shown below:

	Average Years of Secondary and Higher Education of the Population Aged 25-64 in 1976	
	Secondary	Higher
United States	4.75	1.05
12 other countries	3.47	.445
United States as percent of other 12	137	236

Source: Maddison (1982), Table 5.7, p. 110.

The stock of educational experience was larger in the United States than in the other countries, on average, and, in this respect, the United States also ranked high. It was second only to Germany in the stock of secondary education and first, by a wide margin, in higher education.

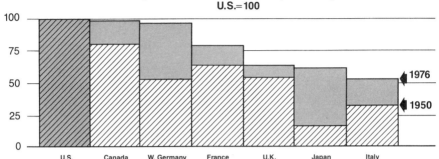

Chart 2-4
Nonresidential Capital Stock per Person Relative to the United States
(Conventional Real Gross Capital Stock)
U.S.=100

Source: Lipsey and Kravis (1987), Appendix Table 10.

Has the United States Misallocated Its Capital Formation?

It is widely believed that the United States not only saves relatively little in conventional terms but also funnels excessively large shares of its saving into residential construction and consumer durables, while other countries channel their saving into forms of investment that are regarded as more productive, such as machinery and equipment. We investigated this question with respect to both conventional capital formation and the broader concepts, as well as with respect to the stock of capital.

The question we can answer is whether there are major differences between the United States and other countries in the allocation of capital. But the question we ideally would wish to answer is whether the allocation of capital has been so distorted by tax or other preferences that the productivity of U.S. capital has been seriously reduced in relation to that in other countries. This is an extremely difficult question and involves, among other things, separating the effects of tax preferences on the allocation of capital by sector of ownership (households vs. business firms) from effects on the type of capital formation (houses vs. factories or cars vs. machinery). Since we do not know of any attempts to perform such an analysis across countries, we will limit the analysis to a consideration of the facts about the allocation of capital among types.

The Distribution of Conventional Capital Formation Among Types of Assets

The share of conventional capital formation going into residential building has not been exceptionally high in the United States. Over the period from 1960 to 1984, half of the countries devoted a higher proportion of their capital formation to residential building (Lipsey and Kravis, 1987, Appendix Table 12).

41

	Residential Building as Percent of Conventional Capital Formation				
	1960-84	1960-69	1970-84	1970-79	1980-84
United States	25.6	26.0	25.4	26.5	22.7
14 other countries	NA	NA	24.8	25.2	24.0
13 other countries, excluding Sweden	24.2	23.2	25.0	25.4	24.1
United States as percent of 14 countries	NA	NA	102	105	95
13 countries	106	119	102	104	94

Source: Lipsey and Kravis (1987), Appendix Table 12.

The share in the United States was about 6 percent above the average of 13 or 14 other developed countries, but it was above the average by a wide margin only in the 1960s and has been below it in the 1980s.

In 1960-84, Denmark, Germany and France invested most heavily in residential building, and Norway least. Japan invested a relatively small share in housing during the 1960s, about 25 percent below the average of the other countries listed, and then moved closer but remained below average in this respect in the 1970s and 1980s (see Lipsey and Kravis, 1987, Appendix Table 12). It should be mentioned that these ratios represent the cost of building and land improvement but not the cost of land, which in Japan is exceptionally high in relation to other housing costs.

To the extent that rapid growth is associated with high levels of outlays for producer durables—machinery and equipment—one might expect to find that the United States has been neglecting this type of capital formation, while the capital formation of the fast-growing countries was heavily tilted in this direction. But, at least for the last 15 years, the data do not bear out this expectation. The share of producer durables, both transport equipment and other machinery and equipment, in conventional capital formation was above average in the United States In fact, it was the United Kingdom, a slow-growth, low-investment country that had the highest share of capital formation going into machinery and equipment, while the share in Japan was below that of the United States and below the average of the other industrial countries!

	Producer Durables as Percent of Conventional Gross Capital Formation 1970-84		
	Total Producer Durables	Machinery and Equipment	Transport Equipment
United States	42.7	32.5	10.2
12 countries, excluding the United States	38.8	29.2	9.6
United States as percent of 12 countries	110	111	106

Source: Lipsey and Kravis (1987), Appendix Table 13.

If we broaden the concept of capital formation to include consumer durables, about half of which are cars, we find some considerable differences in the allocation of investment. The United States led in the proportion of investment going into consumer durables, as we might expect from the much-discussed favorable tax treatment of interest on consumer borrowing, only recently reduced, by the Tax Reform Act of 1986. However, the proportion of capital formation going into consumer durables in Canada was the same as that in the United States, and Denmark, France, the Netherlands, and the United Kingdom were not far behind (Lipsey and Kravis, 1987, Appendix Table 14). The country that devoted very little of its investment to consumer durables was Japan, where the share was under 10 percent, less than half the average of other countries outside the United States. The United States spent most heavily, by far, on "personal transportation equipment," especially in the 1960's, the share being almost twice the average in 9 other countries. Data are not available for Japan, but the ratio must have been low, judging from spending on all consumer durables.

The high expenditure on personal transportation equipment in the United States accounted for all the difference between it and the other countries in the share of durable consumer goods. In other consumer durables, the U.S. share was no higher than that in other countries.

The Distribution of Broadly-Defined Capital Formation

We can look at the distribution of capital formation—both broadly and conventionally defined—for the 1970s and 1980s (Chart 2-5). When we do, it is

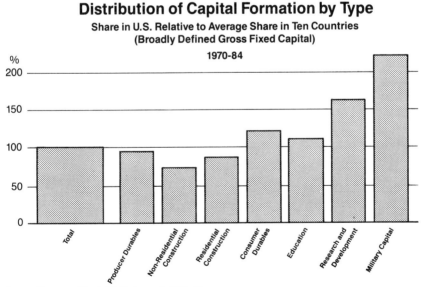

Chart 2-5
Distribution of Capital Formation by Type
Share in U.S. Relative to Average Share in Ten Countries
(Broadly Defined Gross Fixed Capital)
1970-84

Source: Lipsey and Kravis (1987), Appendix Table 15.

clear that conventionally-defined capital formation is a much smaller portion of broadly defined capital formation in the United States than in other countries. That is one reason for the large gap between the United States and others in the usual comparisons. The United States invested a lower share in every conventional form of capital formation and a higher share in each of the nonconventional forms. The United States share in construction was particularly low, even if compared with the share of its capital formation in non-military producer durables. On the other side, the share of U.S. investment going into R&D and military capital formation far exceeded that of other countries.

Shares (%) of Each Type of Investment in Broadly-Defined Capital Formation, 1970-1984

	U.S.	Average of 10 Countries excluding the U.S.	U.S. as percent of average of 10 countries
Broadly-Defined Capital Formation	100.0	100.0	100
Conventional Capital Formation	52.7	62.4	84
Non-Military Construction	30.2	38.4	79
Residential Building	13.4	15.5	87
Other Construction	16.8	22.9	73
Non-Military Producer Durables	22.5	24.0	94
Consumer Durables	19.7	16.4	120
Education	16.6	15.1	110
R&D	6.9	4.2	162
Military Capital Formation	4.2	1.8	226

Source: Lipsey and Kravis (1987), Appendix Table 15.

Japan was the only country with more than three quarters of its capital formation in the types of investment conventionally included in national accounts. It is this concentration, particularly on construction, that places Japan so far above other countries in the usual compilations. The other side of the story is the extremely low share of capital formation going into consumer durables. Japan is also at the bottom of the list in the share of investment going into education, although several other countries are close to it. The heaviest investors in R&D and in military capital formation relative to other forms of investment, were the United States and the United Kingdom; they were also the countries with the lowest shares of capital formation going into construction. The figures in the text table do not indicate any particularly large share of U.S. capital formation or even of U.S. construction expenditures going into residential construction.

As was noted earlier, the United States had the highest share of capital formation applied to consumer durables. However, the shares of some other countries were close to the U.S. level. If the experience of other countries is any criterion, it is difficult to see in these proportions any enormous distortions in the composition of U.S. capital formation from the favorable tax treatment of consumer interest over the 1970 to 1984 period.

The distribution of capital formation is different when it is measured in real values or quantities—that is, translated into a common set of prices for all countries. The translation reduces the share of a type of investment that is particularly expensive in a country and increases the share of a type that is relatively cheap. The effect of using real comparisons for 1975 and 1980 can be seen in the following:

Shares in Broadly-Defined Capital Formation, Nominal Values and at International Prices, 1975 and 1980

	Nominal			Real	
	U.S. (1)	8 Other Countries (2)	U.S. as % of 8 (3)	8 Other Countries (4)	U.S. as % of 8 (5)
1975					
Broadly-Defined Capital Formation	100.0	100.0	100	100.0	100
Conventional Capital Formation	47.9	62.4	77	59.0	81
Non-military construction	27.8	38.3	73	38.1	73
Residential building	9.6	16.5	58	15.7	61
Other building	8.6	12.5	69	9.7	89
Other construction	9.6	9.3	103	12.6	76
Non-military producer durables	20.1	24.1	83	20.9	96
Consumer durables	22.0	16.6	132	19.1	115
Education	18.8	14.3	131	15.8	119
R&D	7.0	4.7	149	4.3	163
Military Capital Formation	4.3	2.0	215	1.8	239
1980					
Broadly-Defined Capital Formation	100.0	100.0	100	100.0	100
Conventional Capital Formation	52.0	59.1	88	58.4	89
Non-military construction	28.5	36.4	78	35.0	81
Residential building	11.0	15.6	71	13.1	84
Other building	8.4	11.9	71	12.5	67
Other construction	9.2	8.8	105	9.4	98
Non-military producer durables	23.4	22.7	103	23.4	100
Consumer durables	20.3	18.3	111	17.3	117
Education	17.2	15.8	109	17.6	98
R&D	6.9	4.7	147	4.6	150
Military Capital Formation	3.6	2.1	171	2.0	180

Source: Lipsey and Kravis (1987), Appendix Table 6e.

With the translation into constant prices, the United States is seen to be closer to the average in the shares of capital formation going into most of the conventional types than the nominal figures suggest, because those types of capital were relatively cheap in the United States. But the shares of non-building construction and education were lower in real terms than in nominal terms because these types of investment were relatively expensive in the United States.

Measured in own-country prices, the share of broadly defined capital formation devoted to construction in 1975 was substantially higher in Japan—

over half—than in other countries (Lipsey and Kravis, 1987, Appendix Table 6e). But measured in world prices, that share was not as high. Construction was very costly in Japan relative to other types of investment. By 1980, the cost of construction was no longer out of line with other Japanese prices and the share of construction in capital formation, nominal or real, was the highest among these countries. The proportion of investment going into non-building construction, mainly infrastructure, was notable in both Japan and Germany, whichever prices were used.

The high proportion of U.S. capital formation allocated to education was noted earlier. The share of education expenses in the United States becomes more like that in other countries when it is measured in world prices, because education was extremely expensive in the United States relative to other forms of investment.

The Distribution of the Stock of Capital Among Types of Assets

Another way to examine the allocation of investment is through the distribution of the stock of capital, using Goldsmith's data for reproducible tangible assets in the late 1970's. The distribution of the stock of capital incorporates not only the cumulation of investment and depreciation over time but also the effects of any differences among the various types of assets with respect to price movements.

The shares of dwellings in reproducible assets, summarized below, reinforce the impressions from shares in capital formation; the data do not give the impression that this category absorbed an exceptional share of investment in

	Shares in Reproducible Tangible Assets Late 1970's	
	Dwellings	Consumer Durables
United States	29.4	11.9
Australia	21.8	4.1
Canada	25.1	9.3
Japan	20.5	5.7
Belgium	34.2	9.5
Denmark	28.1	6.4
France	46.2	6.5
Germany	32.6	10.3
Italy	40.6	9.0
Norway	20.8	3.9
Sweden	31.2	8.3
Switzerland	22.5	4.4
United Kingdom	25.3	9.2
12 countries, excluding United States	29.1	7.2
United States as percent of 12 countries	101	165

Source: Lipsey and Kravis (1987), Appendix Table 9c.

the United States. A few western European countries, in particular, had higher proportions of assets in that form. As might be expected, the share of dwellings was low in Japan, the lowest among the developed countries. It must be kept in mind, especially in connection with Japan, that land is not included in reproducible assets and in national capital formation, although it does enter the capital formation of individual sectors. Land is extremely expensive in Japan and is a large part of tangible assets, over half, as compared with no more than a quarter in any of the other countries recorded (Goldsmith, 1985, Table 37).

Consumer durables play a larger part in U.S. reproducible tangible assets than in those of any other country. These assets were less than half as important in Japan, but there were some European countries where the consumer durables shares were even lower.

We can summarize by saying that the United States seems to be at the upper end of the distribution with respect to investment shares in consumer durables but not in homes, while Japan is at the lower end in both groups. If we think of these two assets as producing services directly for households, we find that France and Italy had about half of their reproducible tangible assets performing this role, and the United States and most European countries had from 35 to 45 percent. At the other extreme, Japan, Australia, Norway, and Switzerland devoted only about a quarter of these assets to such services.

It is important to emphasize that consumer durables are not the only nonhousing assets devoted to providing consumer services. The consumer durables category is distinguished by ownership rather than by function. Countries with low proportions of consumer durables in their assets may have large proportions of similar types of capital under the category of producer durables if they are owned by business firms rather than by households (e.g., consumer durables in rented houses as opposed to those in owner-occupied houses).

Why Do Saving and Capital Formation Rates Differ Among Countries?

Empirical studies of international differences in saving or capital formation are scarce. One reason is that there are only a few countries for which reasonably comparable figures exist, and even for those countries, saving estimates are probably among the least reliable economic data. Secondly, it is necessary to use averages over fairly long periods to remove the influence of cyclical and transitory influences on saving. That means there are few observations on comparative saving levels for sufficient periods of time. Furthermore, there are many factors that have been identified as affecting saving rates, and since they are often strongly correlated with each other, their influence is difficult to sort out with the small number of countries available for study.

There is a good theoretical basis for expecting high rates of income growth to lead to high rates of saving. And most studies of the determinants of saving find that rates of growth of income or productivity are the most consistently significant influences on saving rates (Sturm, 1983). But this relationship points to a problem raised in the next chapter: that it is difficult to distinguish between the effect of saving rates on rates of economic growth and the effect of economic growth rates on saving rates.

Other explanations of differences in saving rates are demographic, relating to the proportion of the population that is working. These include the fraction of the population that is retired and the fraction that is below working age. The higher the proportions of those nonworking groups, the lower the saving rates.

Among the other explanations, particular interest has centered on the effects of social security and similar retirement systems. A higher replacement ratio—the ratio of retirement income to pre-retirement earnings—might tend to reduce saving during the working years, while the expectation of a longer period of retirement, possibly encouraged by the social security system, would tend to raise pre-retirement saving. Even if both effects operated identically in two countries on the rate of saving during working life, the impact on the aggregate saving rate would depend also on the age distribution of the population: the proportion actually in retirement at a given time.

For these, among other reasons, it is not surprising that the measurements of social security effects tend to be somewhat inconclusive and the results erratic. Mostly they are in the expected direction regarding differences among countries. For example, Modigliani and Sterling (1983) found that the below-average income replacement ratio in the Japanese social security system contributed substantially to the above average household saving ratio.

There have been quite a few studies focusing on the apparently huge difference between Japanese and U.S. saving rates. Hayashi (1986), after pointing out that the gap is exaggerated by statistical incomparabilities but that some difference remains after corrections are made, examines most of the popular explanations such as high growth rates, social security arrangements, and demographic factors, and finds them inadequate. He ends by stressing the importance of growth from the low initial level of postwar wealth and, given this circumstance, the desire to make bequests to children. The implication is that the decline in Japanese saving rates since the early 1970s will continue and that they may be expected to approach those of other countries. Ando (1986) stresses rapid growth as the essential explanatory factor.

A common finding of many of these studies, and one we find plausible, is that the saving rate is to a large degree the consequence of factors that, at least over the short run, are not easily manipulated by policymakers. They include the age composition of the population, national growth rates, and the country's real income and wealth levels.

Effects of Taxation

A number of investigators have looked at international differences in the rates at which income from capital and labor are taxed, without producing very conclusive results.

Measurement of the tax burden on capital is a complex problem, since the rate of tax paid varies with rates of inflation, with the sources of finance, the type of assets, sector of ownership, and estimated rates of depreciation. As an example, below are two sets of effective tax rates on capital calculated by Shoven and Tachibanaki:

	Effective Tax Rates on Capital (1980)	
	Zero Inflation	Actual Inflation
United States	32.0	37.2
Japan	29.6	-1.5
Germany	45.1	48.1
Sweden	12.9	35.6
United Kingdom	12.6	3.7

Source: Shoven and Tachibanaki (1985), p. 66. The effective tax rates depend also on depreciation rates and on the proportions of various types of capital. Each country's actual 1980 depreciation rates and distribution by type of capital are used in this calculation.

Neither the rates based on the assumption of zero inflation nor those based on actual inflation give much support to the idea that taxation of capital is a major explanation for differences in the rate of saving. In the zero-inflation calculation, the United States and Japan have almost equal rates, Germany, a relatively high-saving country, has very high tax rates, and Sweden and the United Kingdom, relatively low-saving countries, have low tax rates. With the adjustment for actual inflation, Japanese rates become negative, perhaps helping to explain high saving, but again, tax rates are the highest in Germany, a relatively high-saving country, and extremely low in the United Kingdom, a relatively low-saving country.

Rather different results emerge from a study by Tolley and Shear (1984), based not on tax rates under various circumstances but on an estimated allocation among factors of production of all taxes actually paid. They found the following effective tax rates (percent) on earnings from capital:

	1970	1973	1977
United States	67	62	64
Japan	35	36	29
France	48	50	48
Germany	48	46	47
United Kingdom	62	49	50

By these measures, the United States was the country that taxed earnings on capital most heavily, while Japan taxed them most lightly. While these two observations fit with the idea that relative tax rates were important determinants of investment rates, the virtual equality of U.K., French, and German tax rates does not. The fact that these 1977 rates differ so greatly for the United Kingdom and the United States from the 1980 rates shown above also undermines our confidence in either of these estimates as explanations for differences in capital formation rates.

Differences in taxation have been put forward as an explanation for the gap between U.S. and Japanese saving levels; however, Hayashi (1986) finds "... no strong evidence ... for the effectiveness of the tax incentives for saving." Boskin and Roberts (1986) and Makin (1986) mention the lighter tax treatment of household saving in Japan and the tax preferences for housing in the United States. Yet none of the authors who have studied these differences in taxation appear to believe that they are the major reasons for the gap in saving rates or that they are as important as income growth rates, demographic variables, and retirement practices.

Effects of Government Saving

Our interest in the level of government saving arises from the question of whether individuals take account of it in their saving decisions. Does extravagant government policy dissipate predetermined levels of private saving, or do private savers offset government dissaving by increasing their own, knowing that the government is building up future tax liabilities that they and their children will have to shoulder? Is increased government saving offset by reductions in private saving? If private savers are not influenced by the rate of government saving, differences in government saving rates might account for some of the variance across countries and some of the changes over time in aggregate saving rates.

The idea that private savers ignore government saving or dissaving in their decisions is not easy to contradict or to verify. Private savers might be more conscious of the state of social insurance funds than of that of other parts of the government. Their response may differ from country to country, or from one level of government to another, and over time according to their faith in the government's financial stability. And whether or not there are direct influences on private saving, there are surely indirect ones through effects of government fiscal policy on monetary growth, inflation rates, interest rates, and private income levels. A large government deficit, while it may not directly influence private saving to offset it, may raise private saving by increasing interest rates. Even on this point, however, the size of the effect is uncertain because of the wide range of estimates of the relation between interest rates and saving.

On the whole, the preponderance of empirical evidence for the United States does not appear to support anything like complete offsetting of changes in government saving by changes in private saving. A recent analysis of Italian experience also suggests that changes in private saving do not offset changes in government saving, and that, in fact "... government deficits matter—in that they reduce national saving almost one for one" (Modigliani, Jappelli, and

Pagano, 1985). In view of the evidence that there is not complete offsetting of changes in government saving by changes in private saving, we examine government saving as a possibly separable influence on the aggregate saving rate.

Differences in average government saving rates among countries (government saving relative to GDP) over the period 1970-84 were positively correlated with differences in national saving rates (Appendix Table 2-1); that is, countries in which governments saved more (ran larger surpluses or smaller deficits) had higher national saving rates. There was no relation across countries between government and private saving rates, suggesting that private savers did not offset high or low government saving. Another inference is that government and private saving were not responding to similar forces, such as their country rates of growth.

Still another indication of the effect of government saving on national saving is that countries in which governments increased their dissaving the most between the 1970s and 1980-84 also decreased their saving rates the most (Appendix Table 2-1), while there was no similar relationship between government saving and private saving. In other words, private savers did not offset changes in government saving rates between these two periods. On the face of it then, this evidence suggests that government saving has some independent effect on national saving. Government surpluses or deficits do matter. The importance of government dissaving in the United States in the recent period is put strongly in a statement by Michael Boskin (1985) to the effect that "... federal government dissaving is currently swamping any likely increase in private sector saving that could be produced by structural changes in tax policy in the near future."

A warning must be inserted about these calculations of the relation of government saving to national saving. While we can ignore the role of capital gains and losses, mainly from inflation, in national saving without much distortion, the same cannot be said for measures of sectoral saving such as that of households or particularly governments. Taking account of this effect Eisner calculates U.S. federal net saving as follows, as compared with the OECD measures.

	U.S. Federal Government Net Saving ($ billion)	
	OECD	Eisner
1960-69	−61.3	50.7
1970-79	−273.1	−21.7
1980-83	−347.4	−289.1

Sources: Eisner (1986), Table B-7, p. 192; OECD *National Accounts,* Vol. II.

The difference is not so great in the 1980s, but these calculations give an entirely different picture of the government's role in the 1960s and 1970s, and show an even sharper swing from the earlier position to the heavy federal government dissaving of the 1980s.

A narrower aspect of the relation between public and private saving that has engendered particular controversy in recent years is the effect on private saving rates of Social Security, Medicare, and Medicaid in the United States, and corresponding programs in other countries. If governments balanced their assets with the value of their obligations in these programs, complete substitution of social for private saving would not affect national saving rates. But, that is rarely if ever the case. If households do not take account of the future tax obligations that they or their heirs have to pay for underfinanced social insurance programs, the result should be to reduce aggregate saving. Since living costs and medical costs during retirement are two of the main incentives for household saving, it seems quite plausible that households should take account of the existence of social insurance in making their saving plans. Furthermore, it seems implausible that households are completely aware of the future tax burdens implied by the social insurance funds' obligations, even if they were sufficiently interested in the welfare of future generations to take their liabilities fully into account in saving plans. Thus, the logic of the situation strongly suggests that social insurance plans result in lower national saving.

Unfortunately, the empirical evidence in this area is quite inconclusive, even more so than for the effects on household saving discussed earlier. Sturm (1983) reviewed studies based on individual country time series, cross-country comparisons, and within-country cross-sections, but found the results to be either irrelevant to the determination of effects on aggregate saving or very sensitive to the choice of explanatory variables, time periods, or countries included.

Despite the weak evidence discussed earlier for negative effects of social insurance programs on private saving, we are unable to find even that degree of evidence for effects on national saving. This remains a question on which there is no consensus.

Saving vs. Capital Formation

We have discussed saving and capital formation rates as if they were identical although, as pointed out at the beginning of the chapter, they differ to the extent that a country is financing its capital formation by borrowing from foreigners or is financing foreign capital formation by lending to foreigners. We have been able to do this in analyses of long-run developments for industrial countries, because over periods long enough to average out cyclical changes, the two measures are typically very similar. Simon Kuznets, in discussing trends in the share of capital formation and saving in output, pointed out that the difference was important for only the United Kingdom, among lending countries, and for some of the small borrowing countries in the early stages of development. For the developed countries considered here, the two ratios are very similar, as has been pointed out by Feldstein and Horioka (1980). They reported that for the period 1960-74, the average absolute difference between gross saving and gross capital formation ratios among 21 developed countries was a little over 1 percent of GDP and a little over 5 percent of the saving ratio,

4 percent for the countries we discuss in this chapter. For 1970-79, the difference was about 4 percent of gross capital formation for the 15 countries we cover, but it rose to over seven percent in 1980-84 (Lipsey and Kravis, 1987, Appendix Table 16). Thus for most periods, and particularly for judging changes over longer spans, it is appropriate to use the saving rate and the capital formation rate interchangeably.

The latest period, 1980-84, shows an unusual degree of divergence between the two, and the absolute amounts, particularly for the United States, are enormous. However, even the 1984 U.S. borrowing from abroad in the $100 billion range was only 2.5 percent of GDP, and the borrowing of 1980-84 averaged under 4 percent of gross capital formation. Half of the countries covered here were borrowing from abroad larger portions of the financing of their domestic capital formation in 1980-84 than the United States:

	Gross Capital Formation as Percent of Gross Saving 1980-84
United States	104.0
Australia	121.9
Canada	101.7
Japan	96.8
Austria	103.2
Belgium	117.5
Denmark	123.4
Finland	105.7
France	108.4
Germany	100.4
Italy	107.5
Netherlands	90.0
Norway	88.7
Sweden	112.9
United Kingdom	92.6

Source: Lipsey and Kravis (1987), Appendix Table 16.

The 1984 U.S. borrowing from foreign countries was, of course, much larger relative to gross capital formation—over 13 percent. It is that very recent history and its continuation into the present and probably some time into the future, that provokes justified alarm, but it is important to keep in mind that such foreign financing of capital formation has not been typical of the United States in recent decades.

Summary

We have found that the United States, by a broad definition of capital formation, has been investing a proportion of its gross domestic product over the past fifteen years that is not far below that of other developed countries. Thus, the United States has not been a particularly profligate nation, although most countries save and invest somewhat larger shares of their output.

The U.S. long-run ratio of gross capital formation to GNP, measured by the conventional definition, was about the same before and after World War II. In the earlier period (1869-1938), that ratio placed the United States in the top ranks of countries in this respect, but since 1950 the ratios for other countries have risen to higher levels. While the proportion of output devoted to capital formation may have been a little lower in the United States than in other countries, the United States got more real capital than most countries out of a given amount of investment because the prices of its capital goods were relatively low. Thus, in world prices, or real terms, U.S. investment was higher relative to output than in the nominal terms in which the comparison is usually made.

If we examine real capital formation per capita, even by a narrow definition, the lower U.S. proportion of capital formation in output, combined with the higher U.S. output per capita, resulted in a U.S. investment per capita above that of developed countries as a group. For a broad measure of capital formation, the U.S. margin relative to the average was even larger, and few countries surpassed the United States.

The results of many years of capital formation, as reflected in measures of the capital stock, confirm the impression that the United States continues to be relatively rich in capital. Even by conventional definitions of capital stock, or only slightly expanded definitions, the United States remained, at the beginning of the 1980s, as the country with the largest real capital stock per capita. Although a few small countries may have moved ahead of the United States, the major developed countries, particularly Japan, remained substantially below the U.S. level.

Within the total of capital accumulation the United States did not, as is often supposed, devote an exceptionally large share to residential construction or a particularly small share to machinery and equipment. It did devote a relatively large part of capital formation to consumer durables, however, and again, the real outlier was Japan, in which the share of such investment was very low compared to almost all the other countries. It should be noted that consumer durables are distinguished from producer durables not by their function or by whether they serve the production of consumer goods rather than producer goods but by the fact of their ownership by households. Some part of producer durables in one country may be performing the same functions, under different ownership, as consumer durables in another country.

Taking the broader view of capital formation, we found that Japan devoted an exceptionally large part of total investment to the conventional forms, particularly to construction, and the United States an exceptionally small fraction. That is part of the reason why Japan's capital formation ratio appears so high and that of the United States so low relative to other countries when the comparisons are based on the conventional definitions. The United States spent relatively high proportions of capital formation on consumer durables, military capital formation, R&D, and, to a lesser extent, on education.

Country variations in saving behavior are usually explained in terms of conventional definitions of saving and by saving or capital formation proportions measured in each country's own prices. Therefore, the explanations are not applicable to a good part of our analysis of differences in investment behavior. For what they are worth, however, these analyses point to differences in growth rates as the main source of differences in saving rates, with faster growth associated with higher ratios. There is some much less conclusive evidence that high levels of social security or other pension payments are associated with lower household saving rates.

The level of capital taxation, often thought of as a suitable policy variable for influencing saving rates, has been measured in a consistent way for only a few countries. But the relation to investment rates remains obscure; there is no strong evidence that lower rates produce higher investment, despite the plausibility of that idea.

There does seem to be a preponderance of evidence that the level of government saving or dissaving does affect the level of national saving. In other words, private households and businesses, while they are influenced to some extent by government social security arrangements, do not fully offset changes in government saving by their own saving decisions. The implication of this conclusion supports the view that government deficits really do matter, as opposed to the idea that only the size of government expenditures matters, no matter how they are financed. Large deficits, because they are not offset or not fully offset by private saving, do reduce the national saving rate.

Appendix Table 2.1: General Government and National Gross Saving as Percent of GDP, Conventional Definition of Saving, OECD Data, 14 Countries

	Net Saving as % of GDP								Change in Saving Rate					
	National				Government				National			Government		
	1970-84	1960-69	1970-79	1980-84	1970-84	1960-69	1970-79	1980-84	1970-79 minus 1960-69	1980-84 minus 1970-79		1970-79 minus 1960-69	1980-84 minus 1970-79	
United States	17.88	19.19	18.55	16.56	-1.51[a]	0.79	-0.89	-3.05[b]	-0.64	-1.99		-1.68	-2.16	
Australia	22.45	25.18	23.95	19.44	2.73	NA	3.37	1.44	-1.22	-4.52		NA	-1.93	
Canada	21.37	21.75	21.90	20.30	0.10	2.39	1.39	-2.49	+0.15	-1.60		-0.99	-3.89	
Japan	31.67	35.33	31.99	31.05	3.85	5.46[c]	4.38	2.79	-3.34	-0.94		-5.41	-1.98	
Austria	26.60	27.60	27.82	24.15	4.40	6.37	5.43	2.34	+0.22	-3.67		-0.94	-3.08	
Belgium	20.50	22.61	23.22	15.07	-2.63	1.04	-0.27	-7.34	+0.61	-8.15		-1.31	-7.06	
Finland	25.54	25.43	26.07	24.48	5.43	6.59	6.73	2.84	+0.64	-1.58		+0.14	-3.88	
France	22.65	24.89	24.24	19.46	1.81	4.16	2.68	0.07	-0.65	-4.78		-1.48	-2.61	
Germany	23.27	27.26	24.36	21.09	2.55	5.76	3.33	0.99	-2.90	-3.27		-2.43	-2.35	
Italy	20.37	24.29	21.01	19.10	-4.69[a]	1.29	-4.12	-6.11[b]	-3.34	-0.93		-1.08	-1.58	
Netherlands	23.37	27.28	24.56	21.00	1.40	NA	2.68	-1.17	-2.71	-3.57		NA	-3.86	
Norway	27.26	27.10	26.31	29.17	7.58	7.39[d]	7.30	8.13	-0.79	+2.86		-0.09	+0.82	
Sweden	19.94	25.10	21.76	16.31	4.05	NA	6.59	-1.03	-3.34	-5.45		NA	-7.63	
United Kingdom	18.77	18.83	19.30	17.72	0.32	2.52	1.34	-1.72	+0.47	-1.58		-1.18	-3.06	

[a] 1970-83
[b] 1980-83
[c] 1965-69
[d] 1962-69

Source: OECD *National Accounts*

Chapter 3
The Relation of Capital Formation to Economic Growth

The main reason for public interest in the level of saving and investment is the belief that the rate of investment determines the rate of a country's economic growth. Yet, there is considerable divergence among analysts of economic growth as to the role of capital accumulation. Some of the problems are technical: whether it is the growth of output or the growth of output per capita that is being explained or whether the measure of capital accumulation used is the growth of capital stock, the growth of capital stock per capita, or the proportion of GNP devoted to capital accumulation. But there are also much more fundamental differences that involve basic issues of what causes economic growth.

The Association between Capital Formation and Economic Growth

On one point there is a good deal of agreement: there has been in the past, and continues to be, a fairly strong association between capital accumulation and the growth of output. We have noted both the extraordinary record of economic growth in the period since World War II and the historically high levels of conventionally defined capital formation during that same era. In this section we review the evidence—historical and current—for the existence of this link. To do this on the basis of data for many countries or long time series, we must be content with conventional definitions of saving and capital formation, omitting household capital other than homes, investment in human capital, other forms of investment in intangible assets, such as research and development, and saving that takes the form of real capital gains.

Summarizing the record for the United States, the United Kingdom, Germany, Japan, and Italy, for 1880-1977, Maddison (1982, pp. 54-55) emphasized the close relation of capital growth to aggregate output growth, "... the broadly similar movements in capital and output over the long run, compared with the completely divergent movement in output and labor input. Labour input per capita has declined everywhere to a rather similar degree, but output and capital stock have increased at rates differing widely between countries but similar to each other in each country."

The close link between capital growth and output growth over time suggests the possibility of a similar relationship across countries. If we knew the capital stock of a country, could we predict its real income fairly closely?

An answer to this question, described in the Appendix to this chapter, showed a very strong relationship in 1970 across 50 countries between real capital stock per capita and labor input per capita, on the one hand, and real output per capita on the other, with almost all the explanation provided by the capital stock. A similar test for 1975 for 34 countries, using a completely different set of data, produced the same result and also showed that capital by itself explained more than 95 percent of the differences in real income levels.

The conclusion from these equations is that we can explain per capita income very well across a wide range of countries by the per capita stock of conventionally measured capital. The extremely high correlation between conventional capital per capita and real income per capita almost seems to indicate that there is little room for other determinants of income levels, such as education and other forms of human capital. And in fact, if we add some form of human capital into such an equation, it does not appear to add any explanatory power.

One reason that adding a variable for human capital to an equation based on conventional physical capital does not increase the explanatory power of the equation is that the two are in practice very closely correlated: the countries with more physical capital per capita also have more human capital per worker ($r^2 = .77$). Thus, we cannot easily observe any possible effect of human capital. Despite the lack of explanatory power for education or human capital input in these equations, it is difficult to imagine that it would be efficient to increase greatly the extent and complexity of machinery and equipment to be operated by a poorly educated and poorly trained labor force. In fact, Richard Easterlin has suggested that real per capita income depends mainly on the level of education. Some of the most striking examples of how education enhances the productivity of other factors of production are found in agriculture, where it is relatively easy to relate the characteristics of individual farmers to their choice of technology and their productivity.

There are other ambiguities in these relationships. The extremely close relation between the amount of conventional capital and the level of income may partly reflect the ability of high-income countries to accumulate capital easily. To some extent they may have lots of capital because they have high incomes.

Another comparison based on gross non-residential rather than net total capital stock per capita can be made using Maddison's calculations for 1976. As Chart 3-1 demonstrates, there was a remarkable matching of capital and per capita income for five of six countries relative to the United States in 1976, shown by the closeness of the points to the 45° line. Aside from Germany, which appears to have had a low income level considering its capital stock, capital per capita was very closely related to per capita income. The two indexes were not quite as close in 1950 and 1960, but the relationship between capital and income was still very strong.

Chart 3-1
Relationship Between Capital Stock and Income Seven Countries

Per Capita Non-Residential Capital Stock vs. Real Per Capita Income

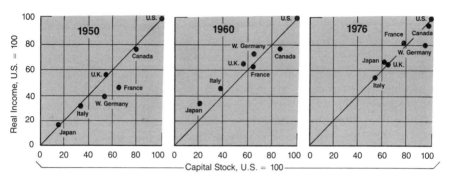

Source: Lipsey and Kravis (1987), Appendix Table 10.

The growth of non-residential capital stock per capita was closely related to the growth in real income per capita over the quarter century from 1950 to 1976 —the ranking of countries by income growth was identical to the ranking by capital growth. However, the relationship was weaker over the two subperiods. In the earlier period, 1950-1960, the growth of output in the four fastest-growing countries, Japan, Germany, Italy, and France, far outstripped the growth of capital. In the second period, however, the growth of capital per capita was much more rapid in three of the same countries, thus catching up with income growth in all of the countries except France.

If we turn to countries other than industrialized market economies, comparisons are most readily available in terms of the ratios of physical investment to gross product. Edward Denison has pointed out that these ratios would not directly explain the growth of output or output per capita, even if the growth of physical capital stock were the crucial determinant of output growth (1967, p. 121). For one thing, the ratio of investment to output is only indirectly related to the growth of the capital stock. Of two countries with the same investment ratio, the one with a lower initial capital stock relative to output will have larger relative additions to capital stock, and the one with a higher price level for capital goods compared with other goods will end up with less capital relative to output. Nevertheless, the investment ratios do reveal something about what might be called comparative effort at capital formation, as mentioned in Chapter 2, and may be worth describing even if their link to economic growth is somewhat tenuous.

We examined the link between investment ratios and the growth of output, total and per capita, for more than 100 countries over five-year periods beginning in 1960. We found that total output growth during each period was greater the higher the average share of GDP going to fixed capital formation and the faster the expansion of the labor force (Appendix Table 3-1). But, the proportion of the variance of growth rates across all countries that was explained by investment in our equations ranged from 13 percent to a little more than a third in developed countries and was lower for developing countries and the world as a whole. In other words, even if the rates of capital formation were affecting the growth rates, rather than the reverse, much of the variation in growth rates remained to be explained by other factors.

The growth of real GDP per capita, which is closer to a measure of consumer welfare than is total GDP growth, was explained to a slightly greater degree by the capital formation ratio and growth of the labor force participation rate— the proportion of the total population that was in the labor force. In this case, variation among developing countries was a little better explained than that among developed countries, mainly because differences in the growth of labor force participation were larger among these countries.

From Capital Formation to Growth or From Growth to Capital Formation?

In view of the tenuous theoretical bases for expecting a strong association between the capital formation ratio and the growth of per capita income, it is not surprising that the correlations are so low. But the looseness of the link between investment ratios and the growth of the real capital stock is not the only ambiguity in the relationship between saving and growth.

Two Schools of Thought

Strong associations between the capital stock and the level of income leave unanswered a question that has long occupied the attention of many analysts of economic development and growth: Does rapid growth lead to high rates of capital formation or do high rates of capital formation lead to rapid growth. We cannot tell from the existence of the relationship between capital and income which way the causation runs or whether it works in both directions. The view—advanced by Walt W. Rostow and Sir Arthur Lewis—that the rate of economic growth depends on the saving rate was referred to by Sen as "... the traditional wisdom of development economics" (1983, p. 750). In its more extreme versions, this view was described by Yotopoulos and Nugent (1976) as "Capital Fundamentalism." The opposite view, strongly expressed by Peter Bauer, is that "... the growth of capital stock cannot explain most of the secular increase in output in the West." Bauer argued that many expenditures justified as investment did not add to productivity and, in fact, were made for political reasons or to expand employment.

In focusing on output per capita, Simon Kuznets and other authors, such as Moses Abramovitz and Edward Denison, did not find capital formation to be the sole determinant of growth or necessarily even the primary one. Kuznets mentioned "... the broad association between modern economic growth and the rise in capital formation proportions for the developed countries..." but seemed to believe that the rise in saving rates was as much a result as a cause of the growth of per capita output. He pointed to some cases of rapid economic growth with low or constant or even declining proportions of capital formation, such as Denmark in the late 19th century. In Norway, Sweden, Canada, and Japan, Kuznets noted capital formation proportions lagging behind growth. He concluded that "Since in several countries high growth rates associated with modern economic growth emerged and were sustained over several decades with relatively low and constant capital formation proportions, we cannot accept unreservedly the notion that high capital formation proportions are *required*" (1973, p. 129). And as Yotopoulos and Nugent (1976, p. 31) summarize the evidence, "... the rise in the investment rate has proven to be neither a necessary nor a sufficient condition for sustained growth."

It is useful in this connection to examine the list of "... widely shared conclusions of policy professionals about the principal 'lessons' associated with successful growth policy" in the concluding chapter of Harberger (1984). There are 13 conclusions, referring to government tax policy, inflation, trade policy, and the role of the public and private sectors, but none on policies specifically directed toward the level of capital formation. That is not to say that the level of capital formation was not considered important in the individual country studies in the Harberger volume. Both the level and allocation of capital formation were frequently mentioned, but they were often treated as having been determined by a broad range of economic policies, those not directed primarily toward capital formation or saving.

Some Empirical Evidence on the Timing of Capital Formation and Growth

If a high rate of capital formation were a prerequisite for rapid economic growth, as is implied in some of the literature on growth, rather than only an accompaniment of it, we might expect to find that increases in GDP, total and per capita, were associated with earlier high levels of capital formation. We therefore examine the relationship of growth in one period to the capital formation ratio of the period five years earlier, together with the contemporary growth of the labor force or of the participation rate. These results do not give much support to the idea that high rates of capital formation must precede high growth rates (Appendix Table 3-2). The capital formation ratios of the preceding period, in combination with growth in labor force or labor force participation, explain each period's growth of real output and real output per capita to a lesser degree than do the contemporary capital formation ratios. In quite a few of the cases, those factors do not explain any visible portion of the country-to-country differences in growth.

The low correlation between output growth and the capital formation ratios of the preceding period suggests the possibility that high rates of capital formation might have been more a consequence than a source of growth. That possibility was tested by relating the growth of total output and output per capita to capital formation in the following period rather than in the preceding period (Appendix Table 3-3). In all six comparisons, growth of output and output per capita is more closely related to the capital formation ratio of the following period than to the ratio of the preceding period. The margin was not as great, but in 4 out of 6 cases the growth of output and output per capita was more closely related to the capital formation ratios of the following period than to the contemporary ratios. On the whole, it appears that there is more evidence that high capital formation ratios follow high growth rates than there is that they precede them or are a requirement for them.

While we do not show equations based on saving rates, we did test that variable as a substitute for capital formation ratios. It was clear that they performed even less effectively than the capital formation ratios in explaining growth in output and output per capita. This is not surprising since it is saving channeled into investment (whether the saving has a domestic or foreign origin) that would promote growth, not saving *per se.*

Has Slow Capital Accumulation Led to Slow U.S. Growth?

Up to this point we have examined the role of capital accumulation in growth in very general terms, without any specific application to the United States. The tentative conclusions from both historical studies and statistical analysis of cross-sectional data for many countries were that, while capital accumulation and growth moved together, it was difficult to say which led and which followed, and on the whole there did not seem to be strong evidence that a high rate of capital accumulation was a precondition for rapid growth. If we had to choose a position, it would be that growth preceded capital formation more often than not.

We now turn to the specific question of whether the low rate of conventional capital accumulation in the United States, relative to other developed countries, was responsible for the relatively low U.S. growth rate.

One way of approaching this question is the growth accounting framework pioneered by Denison and carried on, with some modification, by others. In the growth accounting setting, the contribution of the growth of a factor input—capital, labor or land—to the growth of income is calculated as the product of the growth rate of the factor and the share of the country's total income it receives. Thus, if total output grew over some period from $100 to $200 billion and if capital initially received 20 percent of national income and the amount of capital grew by 50 percent, that growth of capital would be said to have contributed $10 million to the growth in income or 10 percent out of a growth in income of 100 percent. A factor that accounted for a very small share of income, even if it grew very rapidly, could not account for a large contribution to total growth. The way in which this framework is implemented has a strong impact on the resulting estimate of the relative contributions of capital and of the other sources of growth. The definition of output—the use of national income or gross national product—affects the result; the capital shares in contributing to growth will be higher when a gross product measure is used because depreciation is counted as part of gross capital.

Denison's most extensive international comparisons of the sources of growth cover the period after World War II from 1950 to 1962 and are based on the national income concept, averages of gross and net capital, and on a classification of sources of growth into factor inputs and factors affecting output per unit of input (total factor productivity). For the seven countries of Northwest Europe, the growth in national income was about 4.8 percent, as compared with 3.3 percent in the United States (Appendix Table 3-4). Since the United States population was growing at a considerably higher rate than those of the other countries, the United States had the slowest growth in income per capita, and by a substantial margin.

Our interest is in whether this slower U.S. income growth can be accounted for largely by the slower growth of capital in the United States. This method gives little support for the idea that other countries grew faster than the United States mainly because they invested more in capital in the conventional sense. For example, the countries of Northwest Europe grew at an annual rate 1.5 percentage points faster than the United States, but their growth in conventional capital accounted for hardly any (2 percent) of the difference. That is partly because the faster growth of their nonresidential structures and equipment was almost offset by slower growth in dwellings. Even the nonresidential capital growth, however, accounted for only .21 percentage points out of the 1.46 percent faster growth rate. And if we broaden the concept of capital formation to include raising the educational level of the labor force, capital formation was contributing more to growth in the United States than in the other countries, on the average.

A growth accounting framework that departs somewhat from that of Denison has been developed and applied by Christensen, Cummings, and Jorgenson (1980). Their methods and the concepts they use differ from those of Denison in some important respects. Methodologically, for example, they measure the inputs of capital as services on the basis of depreciation and own rates of return, while Denison assumes that a capital stock quantity index provides an index of capital inputs. Conceptually, they work with gross product and rely wholly on gross capital and use the weight of gross capital in calculating the contribution of capital inputs to growth; this contrasts with Denison's use of national income, an average of gross and net to derive the changes in the capital input quantity index and his use of the income share of net capital as a weight. Furthermore, Christensen, Cummings and Jorgenson include consumer durable goods in their measures of capital whereas Denison does not. Other things being equal, their broader concepts of income and capital should produce larger estimates of the contribution of capital to growth.

Their results, despite the major differences in method, agree with Denison's in indicating that differences between the United States and other countries in the growth of capital explained only a small part of differences in the growth of output in the first decade or so after World War II (Appendix Table 3-5). In the next 13 years, however, differences in capital growth were larger and more important, accounting for more than half of the difference in output growth. The higher rate of capital growth in the other seven countries offset a lower rate of labor input growth, so that the growth of total factor input was the same in the United States and the other seven countries. Another way of putting the results is that all the difference in output growth could be accounted for by differences in the growth of total factor productivity.

Conclusions

The issue addressed in this chapter is whether the low rate of accumulation of conventional capital in the United States has been responsible for the relatively low rate of growth of output and output per capita. The facts that point most strongly in this direction are that across countries the size of capital stocks per capita is closely associated with real income per capita and that among countries with very long time series, large increases in capital stock have accompanied large increases in income. The association, strong as it is, does not answer the question of whether the increase in capital brought about the rise of income or the rise of income brought about the increase in capital.

Over shorter periods, the story is somewhat different. Over periods of ten or twenty years, changes in income have not been closely related to changes in capital stock, and the changes in income seem to have preceded those in capital stock as often as the changes in capital stock preceded those in income. Over shorter periods also there was a stronger relationship between the income growth in a period and the rate of capital accumulation five years later than between the capital accumulation and income growth five years later.

Growth accounting analyses of the United States in comparison with other major industrial countries indicate that differences in the rate of capital formation, conventionally defined, accounted for little of the slower growth of the United States during the 1960's, but for a higher proportion in the next decade or so. However, differences in growth across countries seemed to be related more to the growth of combined labor and capital productivity than to differences in the rate of capital formation.

More generally, we accept the view of many analysts of economic development that growth depends on a very large number of factors, of which conventional capital accumulation is only one. Increases in the education of the population and many other factors are too important to neglect in favor of any mechanical relationship between economic growth and capital accumulation.

Appendix Table 3-1: Coefficients for Capital Formation[a] and Growth of Labor Force and Labor Force Participation Rates[b] in Equations for Growth in GDP and GDP per Capita

All Countries
Four Periods

Period	Dependent Variable Growth of:	Independent Variables			$\bar{R}^2$	No. Obs.
		Capital Formation Ratio	Growth of: Labor Force	Growth of: Labor Force Participation Rate		
1965/1960	Real GDP	0.68 (3.76)[c]	1.22 (4.23)		.19	107
1970/1965	Real GDP	1.14 (5.50)	0.91 (3.89)		.25	116
1975/1970	Real GDP	1.14 (4.81)	0.90 (2.84)		.19	117
1980/1975	Real GDP	0.81 (3.60)	0.72 (2.36)		.13	113
1965/1960	Real GDP per Capita	0.60 (3.75)		1.39 (2.55)	.17	107
1970/1965	Real GDP per Capita	1.06 (5.71)		1.18 (3.18)	.28	116
1975/1970	Real GDP per Capita	1.12 (5.29)		0.26 (0.51)	.20	117
1980/1975	Real GDP per Capita	0.68 (3.36)		1.10 (2.79)	.15	113

[a] Gross domestic capital formation as per cent of GDP.
[b] Labor force as per cent of population.
[c] t-statistics in parentheses
Source: IBRD *World Tables*

Appendix Table 3-2: Coefficients for Capital Formation Ratios[a] in Preceding Periods and Growth in Labor Force and Labor Force Participation Rates[b] in Equations for Growth in GDP and GDP per Capita

All Countries
Four Periods

Period	Dependent Variable Growth of:	Independent Variables			$\bar{R}^2$	No. Obs.
		Capital Formation Ratio: Preceding Period	Growth of:			
			Labor Force	Labor Force Participation Rate		
1970/1965	Real GDP	.34 (1.88)[c]	.73 (2.73)		.06	109
1975/1970	Real GDP	.84 (2.78)	.98 (2.95)		.10	116
1980/1975	Real GDP	.25 (1.07)	.83 (2.55)		.04	113
1970/1965	Real GDP per Capita	.38 (2.36)		1.32 (3.10)	.12	109
1975/1970	Real GDP per Capita	.83 (2.99)		.91 (1.64)	.11	116
1980/1975	Real GDP per Capita	.22 (1.07)		1.20 (2.91)	.07	113

[a] Gross domestic capital formation as percent of GDP.
[b] Labor force as per cent of population.
[c] t-statistics in parentheses
Source: IBRD *World Tables*

Appendix Table 3-3: Coefficients for Capital Formation Ratios[a] in Following Periods and Growth of Labor Force and Labor Force Participation Rates[b] in Equations for Growth in GDP and GDP per Capita

All Countries
Four Periods

Period	Dependent Variable Growth of:	Capital Formation Ratio: Following Period	Growth of: Labor Force	Growth of: Labor Force Participation Rate	$\bar{R}^2$	No. Obs.
1965/1960	Real GDP	1.46 (4.30)[c]	2.10 (4.72)		.24	111
1970/1965	Real GDP	.81 (4.66)	.93 (3.88)		.21	115
1975/1970	Real GDP	1.49 (6.43)	.62 (2.08)		.29	114
1965/1960	Real GDP per Capita	1.21 (4.01)		1.19 (1.46)	.16	111
1970/1965	Real GDP per Capita	.75 (4.85)		1.08 (2.80)	.24	115
1975/1970	Real GDP per Capita	1.36 (6.67)		.36 (.77)	.29	114

[a] Gross domestic capital formation as per cent of GDP.
[b] Labor force as per cent of population.
[c] t-statistics in parentheses
Source: IBRD *World Tables*

Appendix Table 3-4: Sources of Economic Growth[a] in the United States and in Northwest Europe[b], 1950-1962

	U.S.	Northwest Europe	Northwest Europe minus U.S.	Contributions as Percent of Growth		
				U.S.	Northwest Europe	Difference
Growth of National Income	3.32	4.78	1.46	100	100	100
Contribution of Growth in						
Total factor input	1.95	1.69	−.26	59	35	−18
Employment	.90	.71	−.19	27	15	−13
Hours of work	−.17	−.14	+.03	−5	−3	2
Age-sex composition	−.10	.03	+.13	−3	−1	9
Conventional capital	.83	.86	+.03	25	18	2
Education	.49	.23	−.26	15	5	−18

[a] Contributions to growth, in percentage points
[b] Belgium, Denmark, France, Germany, Netherlands, Norway, United Kingdom.
Source: Denison (1967), pp. 298, 300.

Appendix Table 3-5: Average Annual Growth Rates of Real Product and Parts Accounted for by Growth of Real Inputs, Nine Countries, Pre-1960 and 1960-1973

	Real Product		Real Capital Input		Real Labor Input		Quality of Hours Worked[f]	
	Pre-1960	1960-1973	Pre-1960	1960-1973	Pre-1960	1960-1973	Pre-1960	1960-1973
United States[a]	3.7	4.3	1.7	1.7	0.6	1.3	0.4	0.5
Canada[a]	5.2	5.1	2.9	2.2	0.7	1.1	0.4	0.3
Japan[b]	8.1	10.9	1.6	4.8	3.1	1.6	0.1	0.4
France[c]	4.9	5.9	1.8	2.6	0.2	0.3	0.3	0.3
Germany[c]	8.2	5.4	2.5	2.8	1.0	−0.4	0.1	0.1
Italy[b]	6.0	4.8	1.3	2.1	0.9	−0.4	0.1	0.8
Netherlands[d]	5.0	5.6	1.9	2.9	0.8	0.2	0.3	0.3
United Kingdom[e]	3.3	3.8	1.7	1.8	0.1	0.0	0.4	NA
Korea	NA	9.7	NA	2.4	NA	3.2	NA	0.8
Average of 7 countries, excluding U.S. and Korea	5.8	5.9	2.0	2.7	1.0	0.3	0.3	0.4
Av. of 7 countries, minus U.S.	2.1	1.6	0.3	1.0	0.4	−1.0	−0.1	−0.1

[a] Pre-1960 = 1947-1960
[b] Pre-1960 = 1952-1960
[c] Pre-1960 = 1950-1960
[d] Pre-1960 = 1951-1960
[e] Pre-1960 = 1955-1960
[f] Included in real labor input

Source: Christensen, Cummings, and Jorgenson (1980)

Chapter 4
Summary

Despite all we have heard in recent years about the low rate of investment and the slow growth of the United States, this country still enjoys a per capita output that is the highest or very close to the highest among the developed countries. Canada and Norway were the countries closest to the United States in 1984, and Germany and Sweden were about 10 percent lower. Japan was not the poorest of the developed countries—Italy and the United Kingdom, among others, were lower—but Japan was far down on the list, with a per capita output about 25 percent below the United States. The most successful and fast-growing developing countries were at around a third or less of the U.S. level.

After World War II, the United States and the United Kingdom grew a little under 2 percent per annum in output per capita and were clearly laggards among the industrial countries (many developing countries were lower). Japan was the leader at around 5¾ percent. The ten EEC countries grew at 3 percent per year, and the average for developing countries was 2⅓ percent, with many exceeding 3 percent.

How do we assess these growth performances? If we look to the 40 or 45 years before World War I or to the period from 1875 to 1950, a growth rate in GDP per capita of more than 2 percent was exceptional and a rate from 1½ to 2 percent was above average. Thus, the period 1950-80 was extraordinary—a golden age of economic growth—judged by historical standards. The growth rate of *world* GDP in these years was substantially above that of the growth of fast-growing countries in earlier periods. Furthermore the growth was widely diffused, including many but, of course, not all, developing countries. As a result, the world has greatly raised its expectations about feasible and even likely rates of growth. One significant manifestation of the rise of expectations was the willingness of U.S. and European commercial banks to lend to developing countries in the 1970s on the basis of long-term growth projections that would have strained credulity in an earlier time.

The growth of U.S. real per capita income in 1950-84 was nearly 2 percent per annum—faster than in the previous 50 years and not far below the rate in 1870-1900. It was higher than the growth rate achieved by any of nine other major industrial countries from 1870 to 1900 and higher than any but Sweden from 1900 to 1925.

Viewed from another perspective, the U.S. growth rate from 1950 to 1984 was close to that for the previous three-quarters of a century; the same was true of Sweden and—to a lesser extent—of Denmark and of the United Kingdom. Thus, the two countries that were less affected by the two World Wars also had the most stable long-term growth rates. (But that is not to say they were the most stable economies in the short run; business cycle fluctuations have typically been relatively wide in the United States.)

The possibility that the post-World War II growth performance includes some degree of making up for past slow growth is most clear for Japan. Per capita output in Japan in 1950 was actually lower than it had been a quarter of a century earlier—the only one of 10 industrial countries for which that was true. In every other case, the average growth of per capita output through the period covering the Great Depression and World War II ranged from ½ percent to 2¾ percent per year, while in Japan per capita output fell at an average rate of about ½ of 1 percent per year. Thus, after World War II Japan was climbing out of a very deep hole. It did not even reach its prewar level of per capita income until the late 1950's and did not recover its prewar standing relative to other countries until much later than that. Nevertheless, the Japanese growth record continued to be outstanding, even after the recovery period.

In conclusion, with respect to the growth of per capita income, the story of the last few decades has been one of many countries reducing the distance by which they lagged behind the United States. Some of the catching up represented recovery from World War II, and much of it represented gaining on the leading economy. But the ability of those countries to sustain such rapid growth rates as they approach or reach the U.S. level remains to be tested.

Differences in Saving and Capital Formation

The United States is widely perceived as a country that saves and invests very little. It is usually described as being at the bottom of the saving league, along with the United Kingdom. It might be 25 percent below the average of other developed countries in the proportion of its income devoted to gross saving and investment and even lower with respect to net saving and investment, when those terms are defined in the conventional way to include non-military construction and equipment. That low rate of saving is blamed for the low rate of U.S. growth compared with other developed countries, and, as a consequence of that assumed "causal connection," a number of policies are being proposed to increase the rate of saving and capital formation.

If, in fact, the U.S. rate of capital formation compares poorly with that of other countries, it is a relatively new situation. Throughout the late 19th century and into the 20th century, the United States had the highest rate of saving and investing or one of the highest, according to most measurements. It is only in the last 35 years that the U.S. rate has been at the low end of the scale. Various explanations have been offered for this decline, including the frequently heard charge that Americans have become present oriented. It is said that individually and as a society, Americans are now much less willing than in the past to give up present consumption for future benefits and future growth.

But this apparent fall to the bottom of the world saving league did not stem from a great change in U.S. habits. What happened is that the share of income going into capital formation rose enormously in the other developed countries and in the developing countries as well, while that of the United States remained close to earlier levels. It was not the United States that changed its ways but the rest of the world.

These conclusions about shares of output going into capital formation are derived mainly from data on what we call "conventional" measures of saving and capital formation. They include the acquisition of structures by all sectors of the economy and of equipment by businesses and governments but not by households. And they exclude not only household acquisition of equipment—autos, washing machines and other durables—but also many other future-oriented expenditures, such as those for education, for vocational training, for research and development, and for military capital goods.

Since we take a broad view, we ask whether the impression of low saving rates in the United States—and the inferred low value placed on future versus present satisfactions—is an artifact of the conventional measurement. We could not calculate measures of saving and capital formation for five or ten countries that would match the broader definitions of saving and capital formation: setting part of current income aside for future use through the acquisition of assets that yield streams of future services. But we did move some distance in that direction, and the effect of such broadening of the concepts is clear. The apparent saving and investment gap between the United States and other developed countries is greatly diminished, even by the limited adjustments we could make. Incorporating expenditures on consumer durables, education, research and development, and military capital formation into national saving and investment reduces the difference between the capital formation rates of the United States and other countries in the period from 1970 to 1984 by about two thirds; the difference declines from 22 percent to eight percent of GDP. Military capital formation plays only a small role in this shift. We believe that the adjustments we could not make for many countries, for such items as the income foregone by students in higher education, would markedly reduce the remaining differences and might even eliminate them.

In other words, we conclude that, as measured by the share of income put aside for future use, Americans are not significantly less forward-looking than people in other countries. Thus, if U.S. economic growth has been relatively slow, it is not primarily because Americans have been unwilling to save for the future.

When we turn from the sacrifice of current consumption to the question of what this foregone consumption purchases, we must take into account the fact that the prices of structures and equipment have typically been lower relative to other goods and services in the United States than in the other developed countries. A given amount of saving buys more long-lived goods in the United

States than in other countries. Taking account of this difference in costs, even without making the adjustments mentioned above, would reduce the difference in saving rates between the United States and other developed countries by several percentage points.

Since the United States remains at the top of the income ladder, another useful point in judging the level of saving and capital formation is that a smaller fraction of income saved can still result in a higher level of capital formation per capita. U.S. capital formation looks more favorable when viewed from that perspective than when seen—as it usually is—in terms of conventionally defined saving rates. Even in terms of conventionally defined real capital formation per capita, the United States level in 1975 and 1980 was above the average of the other developed countries. In 1980 only five of 17 developed countries invested more per capita than the United States. More broadly defined real capital formation per capita in the United States was almost 30 percent above the average for eight other countries in 1975 and was higher than that of any of the other individual countries.

A way of summarizing the relevant history of capital formation is to focus on the capital stock, which would incorporate all of a country's past capital formation, not just that of a recent decade or two, if it were measured correctly. It also would take account of the depreciation or obsolescence of that capital and of any real capital gains or losses from changes of relative prices. Our capital stock measures are not that good. In fact, they are quite imperfect, but they do serve as a check on flows of gross saving and capital formation.

As of 1980, the United States still led its major rivals in capital stock per capita, even by the conventional definition of capital, although some small countries, such as Canada and Norway, appear to have reached higher levels. Thus, the post-WW II period of relatively low levels of conventionally defined U.S. capital formation has left the United States still ahead of the other major countries in conventionally defined capital stock per capita. Other countries, with their higher capital formation ratios, have moved closer to the United States but remain below the U.S. level. The Japanese level continues to be lower, despite Japan's high rate of capital formation, but it is noteworthy that Japan rose from a 1950 per capita capital stock of less than 20 percent that of the United States to about half the U.S. level in 1978. If we use a slightly broader definition of capital stock, one that includes consumer durables, the United States is further in the lead, and Japan is further behind most other countries.

We are not able to calculate values for human capital or for R&D capital for many countries, but such additions would almost certainly place the United States even further ahead of the other developed countries. Some evidence on that point is provided by the substantial margin of the United States over other countries in the average number of years of higher education in the population. This is, to be sure, a crude measure of education capital. There could be great differences from one country to another in the quality of education. Suspicion is often voiced that U.S. education is inferior to that of most European countries; but that judgment is usually based on a comparison of the accomplishments of students at the same level of schooling in different countries. While that may be appropriate for the schooling levels at which education is universal, it is inappropriate for the higher levels at which the United States educates a much larger proportion of its population, possibly reaching down to lower strata of ability than other countries. The proper measure of the output of education is the gain in accomplishment rather than the level reached. It is difficult to compare the quality of given levels of higher education across countries. We can infer, however, from the large number of foreign students in U.S. graduate and professional schools that these schools are highly respected by teachers and students in other countries.

If it is true that the share of U.S. output devoted to broadly defined saving and capital formation is not very different from that of other developed countries, it is natural to look to the composition of capital formation and capital stock to explain the differences in growth rates. It has often been suggested that the composition of capital formation in the United States has been distorted by a tax code that channels inordinate investment into housing. The data on the share of residential building in capital formation, however, do not confirm that conclusion. The U.S. housing share has not been very different from that of most other developed countries. Japan's share has been relatively low, but not far from that of other countries in the 1970s. Thus, if U.S. tax policies favor housing, it may be that similar effects are at work in other countries as well. Or it may be that much of the effect of tax policy is to promote owner-occupancy of housing rather than to increase total investment in it.

Investment in machinery and equipment is often associated with rapid growth, and one might expect to find that share particularly high in Japan. It turns out that this was not the case. In fact, the share of machinery and equipment in total investment was considerably higher in the United States than in Japan during the 1970s.

When we broaden the concept of capital formation to include consumer durables, other major differences emerge. The United States and Canada invested more heavily in this form than any other country, while the Japanese share was extremely low, less than half that of the next lowest country. The high share of consumer durables in U.S. capital formation was entirely accounted for by personal transportation equipment. Americans put no more or even less of their capital formation into furniture and appliances than other countries on average. The fact that the United States and Canada both led in the fraction of capital formation in personally owned cars and trucks suggests that density of population was a factor. However, not all the more sparsely populated countries invested to such a high degree in that category. Another factor may be the extent to which high personal tax rates in some countries offered an inducement to substitute cars provided by employers, which are counted as producer durables, for cars owned by households, which are consumer durables.

An examination of the composition of stocks of reproducible capital, including consumer durables, again fails to show that the United States has invested particularly heavily in housing over the years. Several countries had higher proportions of their capital stock in housing. Japan showed the lowest share of housing in reproducible capital among the developed countries. But the housing measure does not include land, which has been so expensive in Japan that it accounts for over half of total tangible assets, as compared with less than a quarter in all the other countries for which we have data. The low share of housing in Japan may reflect its high price rather than consumer preferences or tax policy.

The United States did have an exceptionally high share of its capital in consumer durables in the late 1970s, almost 12 percent, while Japan's share was only 6 percent; several European countries had even lower shares. The U.S. share was further from the average than the low Japanese share, but Germany was not far behind the United States.

The impression of a high share of U.S. capital formation in housing may stem from a the fact that the United States channels a relatively large part of its capital formation through the household sector—over 50 percent if consumer durables are included. Some European countries (France and Italy) are close in this respect, but in Japan, households account for only a little over a third of gross capital formation. Unless one assumes that governments or business firms provide housing and transportation to households more efficiently than the households do themselves, it is difficult to see why this channeling of investment through households should be deleterious to economic growth. We should not attribute more productivity to appliances in rental housing (producer durables) than to the same appliances in owner-occupied housing (consumer durables) or to company-owned cars (producer durables) than to cars owned by households (consumer durables).

Efforts to explain international differences in the shares of saving or capital formation in national income or output have focused on conventionally defined concepts for the private sector, partly because of a lack of alternative measures. The strongest or most consistent determinants of saving rates appears to be the rate of growth of the economy and the fraction of the population that is employed, but neither one is easy for policymakers to manipulate. It is difficult to find evidence relating to the impact of variables more amenable to policy, although not very amenable, such as taxes on capital or social security arrangements. The studies that have been done do not establish clear and consistent relationships between these policy variables and saving levels. That is not to deny that there are such relationships, but to point out that, so far, the data are inadequate to reveal them unequivocally. At this point, we suspect that differences in taxation are not likely to explain the major differences in saving rates.

The evidence does suggest that national saving rates are affected by government budget or fiscal policy. If changes in government saving levels are not completely offset by changes in private saving, they have an independent effect on national saving. Most studies have concluded that some offsetting occurs, such as private responses to enforced saving through social insurance funds, but that the offsetting is far from complete. Although the conventional measures of government saving and capital formation are even more defective than other conventional saving measures we have discussed, any measure would agree with the conventional ones in showing that the influence of changes in government saving over the last fifteen years has been negative—that is, there has been a shift toward deficits or toward larger deficits—in every one of the countries studied, and declines in government saving have accounted in large part for the decline of national saving rates.

Although we have discussed long-run trends in saving and capital formation as though they were identical, we call attention to the significant differences between them in the 1980s, when several countries were financing their capital formation by borrowing from abroad. The United States was not the outstanding example in terms of the share of capital formation financed in this way, but it did stand out in terms of the absolute amount of foreign borrowing. It should be kept in mind, however, that this is so far a brief and atypical episode in U.S. experience.

Capital Formation and Economic Growth

There is a good deal of discussion about the importance of increasing the rate of U.S. saving that implies that there is a very clear and obvious path from increased capital formation to increased growth. Yet, the literature on economic development ranges from authors who deny there is a strong relationship to those who build models that assume a perfect relationship between the two.

If there is such a relationship, it should be observable in the relationship between per capita capital stock and per capita real income. Across countries, capital stock does seem to be very closely correlated with levels of income. Labor input per capita seems to have no measurable impact at all. Measures of labor skill, which might be thought of as representing some part of the stock of human capital, add almost nothing to the explanation of income per capita once the per capita physical capital stock is included as an explanatory variable. We do not conclude from these facts that human capital has nothing to do with the real income level. The problem is that countries with high levels of education and labor skill also tend to have high levels of physical capital per capita, making it virtually impossible to distinguish the effects of the human capital stock on real income.

Over the quarter century after 1950, in the few countries for which we have data, changes in conventional, non-residential capital per person moved closely with changes in real income per capita, but the relationship was much weaker over 10-year sub-periods. The shorter periods, in fact, pointed to another possibility: that rapid capital stock growth may as much follow rapid income growth as cause it. We analyzed the relationship between capital formation ratios and output growth for a large number of countries over five-year periods from 1960 to 1980. Over these shorter intervals, the capital formation ratio, in combination with the labor force participation rate, explained only a small part of the difference in growth. In fact, the relationship was typically stronger between income growth in one period and capital formation rates in the following period than between income growth in a period and capital formation rates in the preceding period. This finding undermines the idea that a spurt in the capital formation ratio is a necessary prerequisite for growth.

From our calculations and from a review of other studies, we conclude that the interactions between output growth and conventional physical capital formation are too complex to summarize in any simple or one-directional relationship. By and large, output growth does not take place without the growth of the conventional physical capital stock, but the latter is not a necessary antecedent. Furthermore, the more elusive forms of investment, such as in human capital and R&D, may be equally necessary concomitants of rapid economic growth. It would be a mistake to focus exclusively on the conventional physical measures, as is often done.

In summary, we emphasize the continued leading role of the United States as a country at the top of the income distribution and as the major source of new technology. Other countries have had the advantage of absorbing the technology of the leaders and are still able to grow very rapidly as a result, but this advantage diminishes as they approach the levels of the leading countries. It is also a mistake to think of the 1950s and 1960s as the norm for the rate of world growth. These periods were exceptional by historical standards; longer-term rates of growth may be a better standard by which to judge the present.

Looking Ahead

What may be new in the post-WW II years is the increase in the speed with which new technology is diffused and the widening of the range of countries capable of absorbing it. It has been spread partly by the enormous wave of foreign direct investment by firms in leading countries, by exports of advanced machinery, which are of greater and greater importance in world trade, and by the education of students from developing countries in the developed countries. Overall, the ability to absorb technology has been enhanced by an increase in educational levels around the world.

Since the rate of growth of the world economy in the decades following 1950 was far beyond the experience of earlier periods, the prognosis for the future requires some explanation of that change. If the "normal" pace of growth has increased permanently, the United States, growing at a rate similar to that of its past, might find itself left behind by others. On the other hand, if the last few decades witnessed only an improved ability of the followers to catch up to the leader, we might expect a convergence of income and output levels but not necessarily a loss of U.S. leadership.

We suspect that much of the explanation for the surge of world growth and for the rapid accumulation of capital was the spread of more advanced technology. Without the presumption of such a change, it is difficult to account for the high growth of productivity in the countries that experienced rapid growth. The historical evidence and the similarity of the technology used in different countries make it unlikely that the newly applied technology was developed indigenously in each country (Easterlin, 1981). Independent advances were probably largely confined to the small number of already industrialized countries. Developing countries tended to obtain technology by diffusion: through copying, through the training of students, through investment in less developed countries by technologically advanced firms from developed countries, or through deliberate transfers from countries that are technological leaders.

Undoubtedly, some technological changes reach deeply into the structure of society. Easterlin (1981) suggests that the acquisition and application of technology is determined by the extent to which a nation's population has acquired the traits and motivations that go with formal schooling. Independence and nationalism in newly created states produced a determination to advance per capita incomes. A major role was played by technological changes in communications and transportation that brought countries closer together. A part was also played by U.S. policies favoring the diffusion of U.S. technology, motivated partly by a desire to see high levels of foreign production obviate the need for U.S. aid. This stand is unusual if not unique in the annals of public policy, as history is replete with efforts to prevent the outflow of knowledge. For developed countries, the United States organized international groups to analyze and promote productivity in particular industries and arranged for on-site transmission of knowledge by teams of experts. For developing countries, the United States offered technical assistance programs. Direct investment abroad was also encouraged.

Whether owing to the stimulus of U.S. government support or, as is more likely, mainly in response to basic economic incentives, all the major channels for the diffusion of technology were heavily used. Exports, possibly the most important diffusion channel, expanded rapidly, faster than world industrial production. About 60 percent of the manufactured exports of the United States and Japan consisted of machinery and transport equipment in 1984, and the sale of such products often requires producers and distributors to provide technical assistance to their users. Exports often give rise to imitation by producers in importing countries.

Foreign direct investment by U.S. companies also expanded rapidly, especially from the 1950s to the 1970s (Lipsey 1987). The technological gains of the host country from these investments are not limited to the use of new technology; possibly, a larger gain accrues from imitation. The transfer of technology to subsidiaries hastens the speed by which new techniques become known to indigenous producers in the host countries.

Another important factor permitting high foreign growth rates was the opportunity for a redistribution of the labor force. The heavy concentration of the world's labor force in agriculture, which over the centuries has provided a stream of recruits for industry and services, was reduced as the Green Revolution and other technological changes reduced the demand for farm labor. Between 1960 and 1980, the proportion of the labor force engaged in agriculture in "middle-income" countries declined from 62 percent to 46 percent. The change in the industrial countries was from 18 percent to 6 percent (World Development Report 1984). In the United States, the percentage fell from 7 percent to 2 percent. Clearly, there was more leeway for shifts out of agriculture to affect the growth rate in the other industrial countries, and especially the LDCs, than in the United States.

A somewhat different set of explanations for the rapid narrowing of the gap between the United States and other countries in the post World War II period, especially by Germany and Japan, has been offered by Mancur Olson (1982). He stresses the effects of the accumulation of barriers to growth in aging and relatively stable societies—the development of networks of entrenched interests that defend old industries and occupations and the rents from them, resisting or even preventing the innovations that promote growth. Stable societies eventually develop a type of sclerosis, but major events, such as the loss of a war, may sweep away these networks and permit periods of rapid growth. Even less cataclysmic events, such as the formation of the EEC, work partly by breaking up stable national coalitions and by introducing foreign competition.

It is not clear what the implications of this theory are for the issues raised here. As with our discussion in Chapter 3, it suggests that rates of saving and capital formation may be more a concomitant of rapid growth than a cause. A further implication may be that measures to increase the rate of capital formation are less likely to produce major improvements in growth rates than measures to reduce impediments to growth and that such improvements may increase the rates of saving and capital formation.

The idea that there has been a long-term trend toward catching-up to or converging with the level of the leading country or countries is supported in a recent paper by William Baumol (1985), partly based on data from Matthews, Feinstein, and Odling-Smee (1982). These authors focus more on labor productivity levels than on per capita income, but point to many of the same explanations and conclusions. Baumol points to the tendency over the last 100 years or so for productivity to grow more rapidly in the initially low-productivity countries than in the leading countries and to the relative steadiness of U.S. productivity growth. What enforces convergence is the nature of the forces that increase growth. Technological advances have produced an enormous increase of trade and communications since the mid-nineteenth century. Technology is a "public good" insofar as its benign effects are shared everywhere. The impact of these growth-inducing factors, Baumol argues, is so great as to suggest the possibility that "... intercountry differences in growth policy, culture, inventiveness and savings are of minor importance for relative productivity growth" (p. 5).

There are major exceptions to the observation that initially poorer (Chapter 1) or initially less productive (Baumol, 1985) countries grow more rapidly. As we observed earlier, some of the world's poorest countries have not tended to catch up with the rest of the world and have probably fallen further behind. Baumol suggests that one reason is that their economies are so different from those of the advanced countries, particularly in the preponderance of subsistence agriculture. As a result, the scope for absorption of advanced technologies is small because these technologies are applicable only to industries that are absent from, or of negligible importance in, the poorest countries. A different explanation places the blame for slow growth on poor internal policies: taxing primary production to subsidize manufacturing, relying too little on markets and private enterprise (and too much on decision-making by central governments), maintaining overvalued exchange rates, and maintaining a high degree of trade protection.

The United Kingdom is another exception to the catching-up hypothesis. After it lost its lead in labor productivity it showed no tendency to catch up with the countries that had surpassed it, but fell further and further behind.

The apparent contradiction between the virtual inevitability of convergence in the picture drawn by Baumol and the importance of political, cultural, and social factors in Olson's analysis is reconciled to some extent by Moses Abramovitz. He suggests that the initial gap in productivity between a country and the world's leaders defines the potential for growth but that many other elements, including the educational and technical levels of the population and the political and social environment, determine the extent to which a country takes advantage of that potential.

On the whole, our long view of U.S. performance relative to other countries leads to the conclusion that U.S. growth has been quite respectable by historical standards, especially for a leading country. And our examination of the rate at which the United States has been providing for the future, aside from the most recent years, tells us that, while there are many ways of measuring saving and investment, there is no strong evidence that the United States is more improvident than other countries. We doubt that feasible shifts in the rate of accumulation of capital in buildings and equipment would greatly alter the growth performance of the United States relative to the other developed countries. That performance seems to involve much more than any simple relation between growth and physical capital accumulation, and the most useful avenues for capital accumulation may well be in the less tangible types, such as education and research in which the United States has been a leader.

None of this implies that the current or recent past rates of U.S. income growth or capital formation are the highest that can be achieved. The world as a whole has set a new standard of economic growth, and there is no reason to believe that this diffusion of growth and convergence toward U.S. income levels has been damaging to the United States. The leading country does not lead in every field, and as others gain on the United States, their technological advances will spread to the United States just as U.S. technology spread to other countries.

Most of the evidence indicates that many countries have some ability to increase their growth rates, not only through higher physical and intangible forms of investment but also by removing obstacles to growth and preventing the erection of new barriers to growth by groups fearing change. Some of the benefits of policies that encourage growth are confined to the country that initiates them but not all, and each country stands to gain by encouraging others to follow the same path.

While there is no law or single prescription for permanent U.S. leadership, the closest major pursuers still have a way to go to catch up technologically, and they may find the remaining gap harder to close than the initial one. In the meantime, the race is benefiting both the leaders and the pursuers.

List of References

Abramovitz, Moses (1985), "Catching Up and Falling Behind," Paper presented to Meeting of the Economic History Association, September 20.

Adams, F. Gerard and Susan M. Wachter, Eds. (1986), *Savings and Capital Formation: The Policy Options,* Lexington, Mass., Lexington Books.

Alton, T.P., et al (1985), *Economic Growth in Eastern Europe, 1970 and 1975-1984,* New York, L.W. Financial Research.

Ando, Albert, and Arthur Kennickell (1987), "How Much (or Little) Life Cycle is There in Microdata? Cases of U.S. and Japan," in Rudiger Dornbusch, Stanley Fischer, and John Bossons, Eds., *Macroeconomics and Finance: Essays in Honor of Franco Modigliani,* Cambridge, MIT Press.

Auerbach, Alan J. (1985), "Saving in the U.S.: Some Conceptual Issues," in Hendershott, ed. (1985).

Baily, Martin Neil (1981), "The Productivity Growth Slowdown and Capital Accumulation," *American Economic Review,* Vol. 71, No. 2, May, pp. 326-31.

Bauer, Peter (1981), *"Equality, The Third World, and Economic Delusion,* Cambridge, Harvard University Press.

Baumol, William J. (1985), "Productivity Growth, Convergence and Welfare: What the Long Run Data Show," New York University, Department of Economics, C.V. Staff Center for Applied Economics, August.

Berndt, Ernst R. (1984), "Comment," in Kendrick, Ed. (1984).

Blades, Derek W. (1983), "Alternative Measures of Saving," *OECD Occasional Studies,* Paris, OECD.

Blades, Derek W., and Peter Sturm (1982), "The Concept and Measurement of Savings: The United States and Other Industrialized Countries," in *Saving and Government Policy,* Proceedings of a conference held at Melvin Village, New Hampshire, October 1982, sponsored by the Federal Reserve Bank of Boston.

Boskin, Michael (1985), "Theoretical and Empirical Issues in the Measurement, Evaluation, and Interpretation of Post-War U.S. Saving," in Adams and Wachter (1986).

Boskin, Michael J. and John M. Roberts (1986), "A Closer Look at Saving Rates in the United States and Japan," Studies in Fiscal Policy, Working Paper No. 9, Washington, American Enterprise Institute, June.

Brunner, Karl and Allan H. Meltzer (1985), *The New Monetary Economics: Fiscal Issues and Unemployment,* Carnegie-Rochester Conference Series on Public Policy, Vol. 23, Autumn.

China (1984), *Statistical Yearbook of China, 1984*

Christensen, Laurits R., Dianne Cummings, and Dale W. Jorgenson (1980), "Economic Growth, 1947-73; An International Comparison," in John W. Kendrick and Beatrice N. Vaccara, Eds., *New Developments in Productivity Measurement and Analysis,* Studies in Income and Wealth, Vol. 44, University of Chicago Press for NBER.

Darby, Michael (1984), *Labor Force, Employment and Productivity in Historical Perspective,* Los Angeles, University of California, Institute of Industrial Relations.

David, Paul A., and John L. Scadding (1974), "Private Savings: Ultrarationality, Aggregation, and 'Denison's Law,'" *Journal of Political Economy,* Vol. 82, No. 2, Part 1, March, 225-249.

Denison, Edward (1967), *Why Growth Rates Differ,* Washington, Brookings Institution.

—and William K. Chung (1976), *How Japan's Economy Grew So Fast,* Washington, Brookings Institution.

—(1983), "The Interruption of Productivity Growth in the United States," *Economic Journal,* March.

—(1985), *Trends in American Economic Growth, 1929-1982*, Washington, Brookings Institution.

Deutsche Bundesbank (1984), "The Saving Ratio of Households in the Federal Republic of Germany: An International Comparison," in *Monthly Report of the Deutsche Bundesbank,* January.

Easterlin, Richard A. (1981), "Why Isn't the Whole World Developed?" *Journal of Economic History,* XLI.2, March.

Eisner, Robert (1985), "The Total Incomes System of Accounts," *Survey of Current Business,* January, pp. 24-48.

—(1986), *How Real Is the Federal Deficit?,* New York, Macmillan.

Feldstein, Martin, and C. Horioka (1980), "Domestic Savings and International Capital Flows," *Economic Journal,* Vol. 90, No. 358, June.

Giersch, Herbert, and F. Wolter (1983), "Towards an Explanation of the Productivity Slowdown: An Acceleration-Deceleration Hypothesis," *Economic Journal,* March.

Goldsmith, Raymond W. (1985), *Comparative Natural Balance Sheets: A Study of Twenty Countries,* 1688-1978, Chicago, The University of Chicago Press.

Harberger, Arnold C., ed. (1984), *World Economic Growth: Case Studies of Developed and Developing Nations,* San Francisco, ICS Press.

Hayashi, Fumio (1986), "Why is Japan's Saving Rate So Apparently High?" in Stanley Fischer, ed., *NBER Macroeconomics Annual,* 1986, Cambridge, Mass. and London, MIT Press for the National Bureau of Economic Research.

Hendershott, Patric H. and Joe Peek (1985), "Real Household Capital Gains and Wealth Accumulation," in Hendershott, ed. (1985).

Hendershott, Patric H., ed. (1985), *The Level and Composition of Household Saving,* Cambridge, Ballinger.

International Monetary Fund (1985), *International Financial Statistics, Supplement on Exchange Rates,* Supplement Series, No. 9.

Johnson, George (1985), "Investment in and Returns to Education," in Hendershott, ed. (1985).

Jorgenson, Dale (1984), "The Role of Energy in Productivity Growth," in Kendrick, ed. (1984)

Jorgenson, Dale, and Alvaro Pachon (1983), "The Accumulation of Human and Non-Human Capital," in Modigliani and Herring, eds. (1983).

Kendrick, John W. (1976), *Formation and Stocks of Total Capital,* New York, National Bureau of Economic Research.

Kendrick, John W., ed. (1984), *International Comparisons of Productivity and Causes of the Slowdown,* Cambridge, Mass., Ballinger for the American Enterprise Institute.

Kravis, Irving B., Alan Heston, and Robert Summers (1982), *World Product and Income,* Baltimore, Johns Hopkins University Press for the World Bank.

Kravis, Irving B. and Robert E. Lipsey (1984), "The Diffusion of Economic Growth in the World Economy, 1950-80," in Kendrick, ed. (1984).

Kuznets, Simon (1966), *Modern Economic Growth-Rate, Structure, and Spread,* New Haven and London, Yale University Press.

—(1971), *The Economic Growth of Nations,* Cambridge, Harvard University Press.

—(1973), *Population, Capital, and Growth, Selected Essays,* New York, W. W. Norton & Co.

Lindbeck, Asser (1983), "Recent Slowdown of Productivity Growth," *Economic Journal,* March.

Lipsey, Robert E. (1987), "Changing Patterns of International Investment in and by the United States," NBER Working Paper No. 2240, May.

Lipsey, Robert E. and Irving B. Kravis (1987), "Is the U.S. a Spendthrift Nation?", NBER Working Paper, No. 2274, June.

Maddison, Angus (1982), *Phases of Capitalist Development,* Oxford, Oxford University Press.

—(1984) "Comparative Analysis of the Productivity Situation in the Advanced Capitalist Countries," in Kendrick, ed. (1984).

Makin, John H. (1986), "Savings Rates in Japan and the U.S.: The Roles of Tax Policy and Other Factors" in Adams and Wachter (1986).

Matthews, R.C.O., C.H. Feinstein, and J.C. Odling-Smee (1982), *British Economic Growth, 1856-1973*, Stanford, Stanford University Press.

Modigliani, Franco, T. Jappelli, and M. Pagano (1985), "The Impact of Fiscal Policy and Inflation on National Saving: The Italian Case," *Banca Nazionale del Lavoro Quarterly Review*, No. 153, June.

Modigliani, Franco, and Richard Herring, eds. (1983), *The Determinants of National Saving and Wealth*, New York, St. Martin's Press, for the International Economic Association.

Modigliani, Franco and A. Sterling (1983), "Determinants of Private Saving with Special Reference to the Role of Social Security-Cross Country Tests," in Modigliani and Herring, eds. (1983).

OECD, *National Accounts*, various issues, OECD, Paris.

Olson, Mancur (1982), *The Rise and Decline of Nations*, New Haven and London, Yale University Press.

Sen, Amartya (1983), "Development: Which Way Now?", *Economic Journal*, Vol. 93, No. 372, December.

Shoven, John B. and Toshiaki Tachibanaki (1985), "The Taxation of Income from Capital in Japan," Paper at Conference on Research in Income and Wealth, August.

Sturm, Peter H. (1983), "Determinants of Saving: Theory and Evidence," OECD Economic Studies, No. 1, Autumn.

Summers, Robert, Irving B. Kravis, and Alan Heston (1980), "International Comparison of Real Product and its Composition: 1950-1977," *The Review of Income and Wealth*, Series 26, Number 1, March.

Tolley, George S. and William B. Shear (1984), "International Comparison of Tax Rates and their Effects on National Incomes," in Kendrick, ed. (1984).

United Nations (1968), *A System of National Accounts*, Studies in Methods, Series F, No. 2, Rev. 3, New York, Statistical Office of the United Nations.

—(1985), *Demographic Yearbook, 1983*, New York, United Nations.

United Nations and Commission of the European Communities (1986), *World Comparisons of Purchasing Power and Real Product for 1980*, Phase IV of the International Comparison Project (ICP), United Nations and Eurostat.

World Bank (1980), *World Development Report, 1980*.

—(1984a), *World Development Report, 1984*.

—(1984b), *World Tables*.

—(1985), *World Development Report, 1985*.

Yotopoulos, Pan A. and Jeffrey B. Nugent (1976), *Economics of Development: Empirical Investigations*, New York, Harper & Row.